Nanotechnology

Nanotechnology

Global Strategies, Industry Trends and Applications

Edited by

Jurgen Schulte

Asia Pacific Nanotechnology Forum

Telephone (+44) 1243 779777

Email (for orders and customer service enquiries): cs-books@wiley.co.uk
Visit our Home Page on www.wiley.com

Reprinted August 2005

Other Wiley Editorial Offices

John Wiley & Sons Inc., 111 River Street, Hoboken, NJ 07030, USA

Jossey-Bass, 989 Market Street, San Francisco, CA 94103-1741, USA

Wiley-VCH Verlag GmbH, Boschstr. 12, D-69469 Weinheim, Germany

John Wiley & Sons Australia Ltd, 33 Park Road, Milton, Queensland 4064, Australia

John Wiley & Sons (Asia) Pte Ltd, 2 Clementi Loop #02-01, Jin Xing Distripark, Singapore 129809

John Wiley & Sons Canada Ltd, 22 Worcester Road, Etobicoke, Ontario, Canada M9W 1L1

Wiley also publishes its books in a variety of electronic formats. Some content that appears in print may not be available in electronic books.

Library of Congress Cataloging-in-Publication Data
Nanotechnology: global strategies, industry trends & applications/edited by Jurgen Schulte.
p. cm.
Includes bibliographical references and index.
ISBN 0-470-85400-6 (cloth : alk. paper)
1. Nanotechnology. I. Schulte, Jurgen.
T174.7.N3738 2005
620′.5–dc22

2004023729
British Library Cataloguing in Publication Data

A catalogue record for this book is available from the British Library

ISBN 10: 0-470-85400-6 (HB) ISBN 13: 978-0-470-85400-6 (HB)

Typeset in 10/12pt Times Roman by Thomson Press (India) Limited, New Delhi
Printed and bound in Great Britain by TJ International, Padstow, Cornwall
This book is printed on acid-free paper responsibly manufactured from sustainable forestry in which at least two trees are planted for each one used for paper production.

Contents

List of Contributors

Jurgen Schulte
Asia Pacific Nanotechnology Forum, The Meriton Heritage Building, Suite 1, Level 2, Kent Street 533-539, Sydney NSW 2000, Australia schulte@apnf.org

Hongchen Gu
Engineering Research Center for Nanoscience and Technology, Shanghai Jiaotong University, Shanghai 20030, China hcgu@sjtu.edu.cn

Wonbong Choi
Florida International University, Mechanical & Materials Engineering, College of Engineering, Engineering Center 3465, 10555, W. Flagler Street, Miami, United States FL 33174 choiw@fiu.edu

Jo-Won Lee
The National Program for Tera-level Nanodevices, 39-1 Hawolgok-dong, Sungbuk-ku, Seoul 136-791, Korea jwlee@nanotech.re.kr

Ottilia Saxl
Institute of Nanotechnology, 6 The Alpha Centre, Stirling University Innovation Park, Stirling FK9 4NF, United Kingdom o.saxl@nano.org.uk

M. C. Roco
Chair of the Subcommittee on Nanoscale Science, National Science Foundation, 4201 Wilson Boulevard, Suite 505, Arlington, Virginia 22230, United States mroco@nsf.gov

Po Chi Wu
92 Melody Lane, Orinda, CA 94563 United States pochiwu@earthlink.net

Tsuneo Nakahara
Sumitomo Electric Industries Ltd, 1-3-12 Motoakasaka, Minato-ku, Tokyo 107-8468, Japan nakahara-tsuneo@sei.co.jp

Takahiro Imai
Semiconductor R&D Laboratories, Sumitomo Electric Industries Ltd, 1-1-1 Koy-akita, Itami, Hyogo 664-0016, Japan imai-takahiro@sei.co.jp

Teik-Cheng Lim
Nanoscience and Nanotechnology Initiative, Faculty of Engineering, 9 Engineering Drive 1, National University of Singapore, S 117576, Republic of Singapore alan_tc_lim@yahoo.com

Seeram Ramakrishna
Dean's Office, Faculty of Engineering, 9 Engineering Drive 1, National University of Singapore, S 117576, Republic of Singapore seeram@nus.edu.sg

David Soane
Nano-Tex LLC, 5770 Shellmound Street, Emeryville, CA 94608, United States soane.david@nano-tex.com

David Offord
Nano-Tex LLC, 5770 Shellmound Street, Emeryville, CA 94608, United States offord.david@nano-tex.com

William Ware
Nano-Tex LLC, 5770 Shellmound Street, Emeryville, CA 94608, United States ware.bill@nano-tex.com

Isao Kojima
Head of the Materials Characterization Division, Metrology Institute of Japan, National Institute of Advanced Industrial Science and Technology, AIST Tsukuba Central 5, Tsukuba, Ibaraki 305-8565, Japan i.kojima@aist.go.jp

Tetsuya Baba
Head of the Material Properties and Metrological Statistics Division, Metrology Institute of Japan, National Institute of Advanced Industrial Science and Technology, AIST Tsukuba Central 3, Tsukuba, Ibaraki 305-8563, Japan t.baba@aist.go.jp

Foreword

In April 2000, the Japanese government established the National Strategy for Industrial Technology in order to identify challenges and solutions for Japanese industrial technology in the twenty-first century. The Second Science and Technology Basic Plan, a five year plan that started in 2001, is a part of this national strategy. According to this plan, a total of approximately $200 billion will be invested in governmental research and development. One of the most significant policies of concern is the prioritization of research and development based on pressing national and social issues in areas such as life science, IT, environment, and nanotechnology and nanomaterials. Nanotechnology is expected to be a key technology underlying a wide range of industrial fields such as IT, energy, biotechnology, and medicine.

Japan's efforts in nanoscience and nanotechnology were initiated by the Atom Technology Project, a ten-year endeavor that started in 1992 and was sponsored by the Ministry of Economy, Trade and Industry (METI) and managed by a quasi-governmental organization which included METI's national institutes. The National Institute of Advanced Industrial Science and Technology (AIST), which was established by the reorganization of METI's national institutes, strives to promote research in nanotechnology based on the results of the Atom Technology Project.

Since nanotechnology is a precompetitive, interdisciplinary, and comprehensive research field, a global network is essential for the further promotion of research activities from nanotechnology to nano-industry. In addition, it is necessary to establish strong regional coordination in order to promote the strategies of the Asia Pacific region to the global standard.

We sincerely hope the Asia Pacific Nanotechnology Forum (APNF) will play the role of catalyst among the Asia Pacific countries.

Professor Hiroyuki Yoshikawa

President,
National Institute of Advanced Industrial Science and Technology

Introduction: Movements in Nanotechnology

Jurgen Schulte
Asia Pacific Nanotechnology Forum

Nanotechnology entered the more public arena in 2001 when President Clinton brought worldwide attention to nanotechnology through his budget approval for the US National Nanotechnology Initiative (NNI). The initial budget allocated for nanotechnology in 2001 was $422 million, which demonstrated the anticipated relevance of nanotechnology to the USA economic growth as well as nanotechnology's strategic importance to national security. Three years later, in December 2003, President Bush signed the 21st Century Nanotechnology Research and Development Act, which allocated a budget of $849 million to the NNI, doubling the initial budget from 2001. After the clear message of commitment of the United States to nanotechnology since 2001, governments around the world reassessed their current national nanotechnology policies or finally began to develop their own focused long-term position in nanotechnology. Since then national government investments into nanotechnology have increased to over $3 billion worldwide in 2003. A number of state governments have started to implement their nanotechnology support in addition to the already existing, substantial national government funding, bringing the public funding for nanotechnology up to an estimated total of $4 billion. Investment by industry alone is estimated to have added another $1 billion in 2003 (APNF).

What is it that nanotechnology to offer that has prompted governments and industry alike to commit substantial funds to a technology which science geeks

Nanotechnology: Global Strategies, Industry Trends and Applications Edited by J. Schulte
 ISBN: 0-470-85400-6 (HB)

started to talk about only a few years ago? What is nanotechnology bringing to the innovation table that cannot be ignored by financial and economic planners? Nanotechnology aims to master engineering at the nanometre scale. Nanotechnology deals with the engineering at a size scale that is currently challenging the entire semiconductor industry in its effort to further miniaturize chip design. It is working at a scale which is so familiar and natural to the biotechnology industry since its own inception. What sets nanotechnology aside from biotechnology is that nanotechnology in addition to biological building blocks also deals with the engineering and controlling of building blocks of any inorganic as well as mixed biological/inorganic building blocks at the nano-scale (natural and/or artificial origin). Mastering of technology at this size scale has the potential of being able to customize any thinkable type of material the way we want it to be or to create properties which only a few years ago were thought not to be in reach in the near future.

Will this then introduce a new generation of gadgets more powerful and more versatile than the ones we have just bought? The answer is yes, but it is only a partial yes since the (electronic) gadgets industry will take some time (6–10 years) until it will have mastered the mass production of nano-engineered based components. Nanotechnology will have a more profound and immediate impact on industry dealing with materials such as aerospace, automobiles, coatings, construction, cosmetics, ceramics, composites, agriculture, detergents, die moulding, drug delivery, fertilizers, food, fuel production, lubricants, medical supplies, metals, optical equipment, paint, paper, pharmaceuticals, polymers, power generation, sensors, tools and textiles, to name but only a few industries that come to mind immediately. The rapid pace at which nanotechnology is moving forward is probably the main reason that it has come to the attention of almost every policy maker and senior manager.

Although biotechnology has long been observed and supported, its full commercial potential is expected to come to fruition only within the next 20 years or so. In contrast to biotechnology, nanotechnology has already found its application in some of the largest industries, such as the textile industry, due to the advantage of not requiring the long clinical trials that are needed for any new biotechnology product. Another major difference between biotechnology and nanotechnology is that biotechnology focuses on a single main base material, i.e. genes and cells, while engineering at the nanoscale is almost unlimited in its choice of base materials, may it be mineral, plant, animal, human, or combinations of such. That, of course, raises some brows. Consequently, nanotechnology has found an immediate spot on agendas of policy makes and industry leaders.

Figure 0.1 shows the public national funding over the past seven years as estimated by the US National Science Foundation (NSF). Since 2001 an average of about 20–30% of national funding is matched by state (province) funding. Private funding (industry, venture capital) is estimated to be larger than government funding except for the smaller Asian development countries. After an initial boost by governments, funding for nanotechnology has steadily decreased over the past

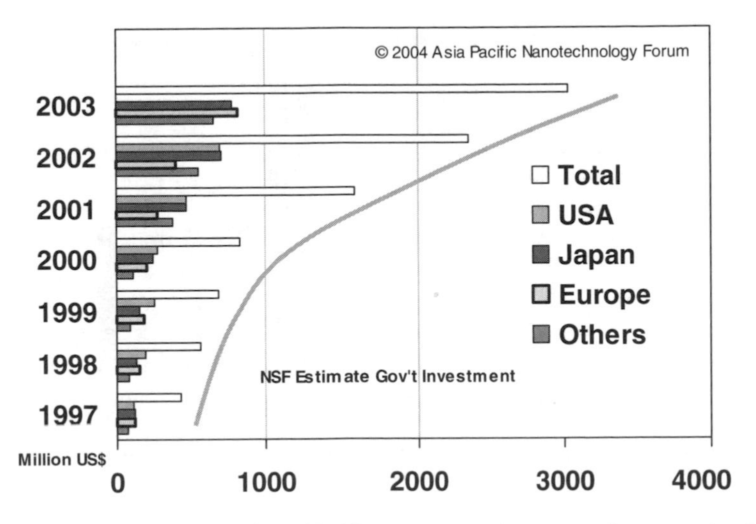

Figure 0.1 NSF estimates of worldwide government investment in nanotechnology

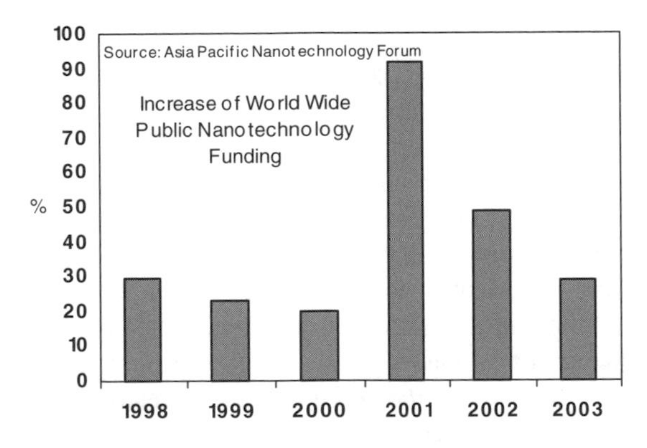

Figure 0.2 Increase in worldwide government investment in nanotechnology

years, which reflects both cautious national budgeting and the additional financial input from private investment as the first R&D spin-offs and products are entering the market (Figure 0.2). Interestingly, this general trend is not shared by Europe and Asia (excluding Japan). In Europe, additional public investment in nanotechnology is increasing at a staggering rate (Figure 0.3).

Most of Asia is used to long-term national plans raging from 5 to 10 years, which requires considerable planning ahead, including major infrastructure building. China's investment commitment for the five-year plan ending in 2005 is $280 million;

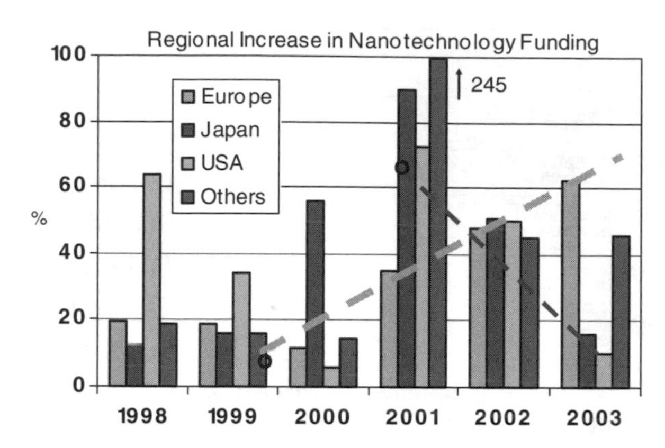

Figure 0.3 Increase in government nanotechnology funding by region. The funding for nanotechnology in Western Europe alone increased by 245% in 2001

for Korea's ten-year plan ending in 2010 it is \$2 billion; and it is \$620 million for Taiwan's five-year plan ending in 2007. Malaysia allocated 9% (\$23 million) of its 8th five-year plan to nanotechnology and precision engineering (Plan Intensification of Priority Research Areas) ending 2008. Thailand is earmarking \$25 million for nanotechnology for the five-year period ending in 2008, and Japan will be investing over \$900 million each year over the next five years. Australia has identified nanotechnology as one of its four funding areas of national priority.

Asia's public investment in nanotechnology is now surpassing the public investment in nanotechnology of all its western competitor combined. The absolute figures indicate how important it has become to take a stake in nanotechnology early on and how important it has become to secure fundamental, core intellectual property to secure a future market share in nanotechnology. Comparing those figures it is important to note that the dollar value of investment in Asia usually has a much better return of investment per dollar than its counterpart in the USA and Europe due to the relatively low cost of highly skilled labour, considerable tax concessions, and ready access to a very large manufacturing industry.

The first movement in nanotechnology has started. Similar to what has been observed in the booming IT industry, this first movement in nanotechnology may decide the future market share and market focus.

Part One

National Nanotechnology Initiatives in Asia, Europe and the US

1

Scientific Development and Industrial Application of Nanotechnology in China

Hongchen Gu[1] and Jurgen Schulte[2]
[1]Nanotechnology and Nanoengineering Center, Shanghai Jiaotong University and [2]Asia Pacific Nanotechnology Forum

With the recent release of a five-year plan for the strategic development of nanotechnology in China, the People's Republic of China has set the pace in nanotechnology development. This chapter summarizes the current status of nanotechnology in China and policies that have been set in place.

1.1 Policy and Objective of Nanotechnology Development in China

In consultation with the National Development and Program Committee, the Ministry of Education, the Chinese Academy of Sciences and the National Natural Science Foundation Committee (NNSFC), in July 2001 the Ministry of Science and Technology issued a policy plan for a national nanotechnology development strategy for the period 2001 to 2010. This draft plan confirmed the general strategy and objective of nanotechnology development in China.

Nanotechnology: Global Strategies, Industry Trends and Applications Edited by J. Schulte
 ISBN: 0-470-85400-6 (HB)

1.1.1 General Strategy

According to the policy plan, the Chinese government is committed to continuously improve innovative capability, develop advanced technology, and finally attain industrial applications relevant to China's present status with a focus on national long-term development. With this plan the Chinese government also made clear that it will insist on its set principle that it will support what is beneficial to China, i.e. catching up with international development in general, while finding breakthroughs that can solve key problems in China. In basic research and advanced technology, exploration and innovation are emphasized; In applications, the development of nanomaterials is the main objective for the near future. The development of bionanotechnology and nanomedical technology is a main objective for the medium term, whereas the development of nanoelectronics and nanochips is a long-term objective. The draft emphasizes that developments in identified key areas need to be well coordinated across departments and disciplines, and a well-structured intellectual property portfolio has to be developed.

The tenth five-year plan emphasizes

- enhancing basic and applied research in nanotechnology;
- exploring possible technology applications depending on market requirement and in line with national development objectives, and promoting the industrialization of nanotechnology with a focus on mass production, education and research;
- establishing a nanotechnology centre and progressively forming an innovative national nanotechnology system.

The key tasks for nanotechnology development in China are

- to align R&D with market requirement;
- to accelerate multidisciplinary R&D and communication;
- to pay particular attention to intellectual property rights and to encourage combined fundamental and applied research, and to pay particular attention to intellectual property rights;
- to align innovation policy with nanotechnology development.

1.1.2 Research Objectives within the Tenth Five-Year Plan

Fundamental research into nanotechnology focuses on the basic principles of physical and chemical characteristics at the nanoscale with the purpose of finding new concepts and new theories. Examples are the development of innovative nanochips, new quantum configurations and new quantum domino effects. Further targets are the physical, chemical and biological characterization of materials at the nanoscale, and the characterization of single molecules and their interaction. The knowledge that will be acquired through fundamental research will provide the basis for the development of advanced scientific theories for the design and manufacturing of new nanostructures, nanomaterials and nanochips based on atomic and molecular technology. The fundamental research is expected to explain phenomena and characteristics at the nanoscale. A further important part of the tenth five-year plan is the establishment of a corresponding nanotechnology database, a national standard for nanoscale, and processes for the industrialization of nanotechnology.

1.2 Application of Nanotechnology

1.2.1 Materials Processing

Nanotechnology is expected to enable environmentally friendly mass production at low cost. It is also expected that nanotechnology will prove to be useful in the development of light and strong materials, biomedical materials, pharmaceutical materials and multifunctional intelligent materials.

1.2.2 Nanochip Fabrication and Integration

Nanotechnology can produce stable and reproducible atomic manipulation plus spontaneous growth; it can deliver super high density memory technology plus integration and encapsulation in nanochip technology. The development of multifunctional nanochips with high integration would offer considerable improvements in speed performance, storage density and power consumption over present systems.

1.2.3 Nanochip Processing Methods

By combining top-down and bottom-up nanoprocessing technology, using microbeam processing and etching technology, and physical, chemical and biological methods of periodic nanopatterning, it is planned to develop nanoelectromechanical systems (NEMS) and optical signal processing systems as well as optoelectric devices.

1.3 Analysis and Characterization of Structure and Function

It is recognized that through scanning tunnelling microscopy and three-dimensional measurements, nanotechnology enables the characterization of single molecules and nanostructures as well as biochemical reactions in cells.

1.3.1 National Safety

Nanotechnology is expected to contribute to China's defence efficiency and capability through development of purpose-designed nanomaterials, functionalized special-purpose materials, nanosensors, micro-engine technology, micro and nano aircraft, and special-purpose satellites.

1.3.2 Technology Transfer and Applications of Nanomaterials and Nanochips

Technology transfer and the application of nanotechnology is promoted through collaboration and amalgamation with advanced technologies and in combination with traditional technologies. During the tenth five-year plan, attention is focused on the development and application of nanotechnology in new materials, computer and information systems, energy, environment, medical, hygiene, biology, agriculture and traditional industry.

The tenth five-year plan has a particular focus on developing nanomaterials technology, which has a beneficial impact on national economics and safety. The research focus in nanostructure material is on the development of heat-resistant materials of high strength and light mass, which can be applied to space navigation and traffic. With respect to research into nanofunctional materials, it emphasizes on the exploitation, preparation and processing of innovative nanomaterials with application in information technology, communication, medical treatment, public health and environment. The abundant natural resources in China provide a basis for the development of innovative nanostructural and nanofunctional materials.

Further areas are the development of nanocatalysts, detergents and combustion-supporting agents for improving the efficiency of traditional energy sources; developing nanotechnology for air decontamination and water treatment; developing technology that can improve the usage of traditional energy sources by greatly decreasing unwanted combustion products. Research into possible negative environmental effects of nanotechnology are also included as well as programmes to enhance and encourage the application of nanotechnology in basic industries, such as chemical engineering, construction materials and the textile industry, to accelerate evaluation and transformation.

Further attention will be given to the speed-up of cross-disciplinary work and amalgamation of nanotechnology with biotechnology, biomedical engineering and traditional medical technology. The aims are to develop nanotechnology for biological detection, diagnosis, treatment and medication; to prepare highly efficient nanomedicine; to improve disease diagnostics and treatment; to develop technologies for better plant disease resistance, insect pest resistance and flexibility to the environment; and to enhance agricultural yield.

During the tenth five-year plan, the Chinese government will increase support for industrialization of nanotechnology, foster corporations with advanced technology, and build an industrial basis. Close collaboration between government, universities, research organizations and industry will be supported to bring scientists, technologists, administrators, industrialists and financiers together, as well as to integrate technology, industrial capital and financial capital, and to combine industrial mechanism and risk investment mechanism to accelerate the industrialization of nanotechnology and economic growth through nanotechnology.

1.3.3 Building Basic Nanotechnology Centres and an R&D Base

According to China's internal competition principle, it is planned to select several national laboratories and related research bases from present laboratories and bases and give them strong support so that they may become the key laboratories for nanotechnology development in China. Here are two specific actions:

- Establish a national science research centre for nanotechnology with advanced instrumentation and state-of-the-art equipment to enable it to become the national pilot centre for nanotechnology, the designing and manufacturing centre for nanochip development, and an R&D centre for nanotechnology. The centre will collaborate with other national laboratories to form a larger network in China. The centre will be open and flexible to integrate excellent scientists from all over China, and to encourage a multidisciplinary environment.
- Establish a national engineering research centre for nanotechnology and its applications to accelerate the innovation of nanotechnology and its industrialization. This centre will combine manufacturing, teaching and research, development of intellectual property, and innovative technology and products, to form a good mechanism for R&D and industrialization. The centre is expected to establish a nanotechnology network, build a nanotechnology information desk, and accelerate the sharing of information sources. It is also expected to encourage corporations to participate in nanotechnology development, and to unite government departments to establish laboratories and engineering research centres.

1.4 Main Policy and Measures

1.4.1 Enhance Leadership and Coordination of Nanotechnology R&D

The central government established the Guidance and Coordination Committee of National Nanotechnology to oversee the national nanotechnology development and to guide and coordinate nanotechnology tasks. The committee consists of the National Plan Committee, the National Economic and Trading Committee, the Ministry of Education, Science and Technology Committee, the National Defense Committee, the Ministry of Finance, the Chinese Academy of Science, the Chinese Academy of Engineering, the National Scientific Foundation Committee and the Ministry of Liberation Army General Supply. The committee secretariat was appointed by the Ministry of Science and Technology.

1.4.2 Implementation of National Nanotechnology Initiatives

According to the task in the compendium, respective resources are to be amalgamaed in an effort to implement national nanotechnology initiatives as outlined in the tenth five-year plan, and to deploy and coordinate the development of nanotechnology nationally. Initiatives are overseen by the National Nanotechnology Committee, supported by the National Scientific Foundation Committee, the National Program for Basic Study and Development, the National Research Plan for Advanced Technology, the National Technology Plan for Solving Key Problems, and the Project of Innovating Knowledge and Education Development Plan Facing the 21st Century. The initiatives are then brought into effect by the respective government organizations and development agencies. Sufficient funding will be made available to support initiatives, human resources and management.

The Chinese government is committed to fostering nanotechnology development, to expediting the construction of national nanotechnology centres and bases, to furthering organization and implementation of basic studies into nanotechnology and innovative advanced technology, and to promoting and fostering individuals with excellent abilities. Respective departments and local governments are directed to confirm objectives and tasks on the basis of the task compendium, to support the implementation of national nanotechnology initiatives.

1.4.3 Encourage All Participants and Create Environmentally Beneficial Nanotechnology

As a first step within the larger initiatives, the National Science Research Center for Nanotechnology and the National Engineering Research Center for Nanotechnology and Application are being built in close collaboration with local government. A link between technology, commercialization and economic growth will help to encourage corporations to participate in the development of nanotechnology in the near future. The Chinese are whole-heartedly committed to supporting technology transfer of nanotechnology development and its industrialization through the frameworks of the National Torch Plan, the New Products Plan, the Technology Innovation Fund for Medium and Small Corporations, the Development Plan of Industries with Advanced Technologies, the Developing Economic Plan Based on Advanced Technology and the Technique Reconstruction Plan.

1.4.4 Foster Scientific Specialists and Technologists in Nanotechnology

As part of the initiatives, scientific specialists and technologists will be supported and international experts will be attracted to meet the demand of specialist researchers in all focus areas. The long-term support of human resources in nanotechnology is addressed by popularizing nanotechnology knowledge in primary and secondary schools to ensure that nanotechnology becomes widely understood. In addition, new nanotechnology disciplines will be created in schools plus new nanotechnology-related courses in physics, chemistry, biology, mechanics, economics and computer sciences.

1.5 Status of Nanotechnology Research in China

Since the Compendium of National Nanotechnology was carried out, the Chinese government has focused on the study of nanomaterials and nanotechnology. National initiatives and local government initiatives invested funds through the National Technology Plan to Tackle Key Problems, programmes 863 and 973, to enable the development of competitive Chinese nanotechnology research publications of a large number of research achievements, and the development of intellectual property, which attracted wide international attention.

1.6 Distribution of Research Potential

1.6.1 Geographic Distribution

There are two main R&D centres for nanomaterials and nanotechnology in China, the northern centre and the southern centre.

The Northern Nanotechnology Research Center is located in Beijing and it includes the Nanotechnology Center and the Institutes of Chemistry, Physics, Metallurgy and Semiconductors within the Chinese Academy of Science, the Beijing Institute of Construction Materials Research, the Beijing Steel Chief Research Institute, Beijing University, Tsinghua University, Beijing Science and Technology University, Beijing Chemical Engincering University, Beijing Science and Engineering University, Beijing Normal University, Tianjing University, Nankai University and Jilin University.

The Southern Nanotechnology Research Center is located in Shanghai and it includes the Shanghai Jiaotong University, Huadong Science and Engineering University, Fudan University, Huadong Normal University, Tongji University, Chinese Science and Technology University, Zhejiang University, Nanjing University, Shandong University, the Institutes of Solid Physics, Metallurgy, Silicates, and Nuclear Science, and the Shanghai Technological Physics Institute within the Chinese Academy of Science.

Apart from these two main R&D centres, nanotechnology and nanomaterials research is also concentrated in the cities of Xian and Lanzhou in the north-west, Chengdu in the south-west and Wuhan in the south.

The geographic distribution of nanotechnology development indicates that the research potential in nanotechnology is spread all over China, but is mainly focused on the areas of Huadong and Huabei, which account for 80% of the overall distribution (Figure 1.1). The survey also indicates that the distribution of research potential seems to be concentrated very locally, but in fact it is spread over a much larger area. For example, the southern R&D centre is mainly located in Shanghai, but also spreads around the cities of Hefei, Nanjing, etc.

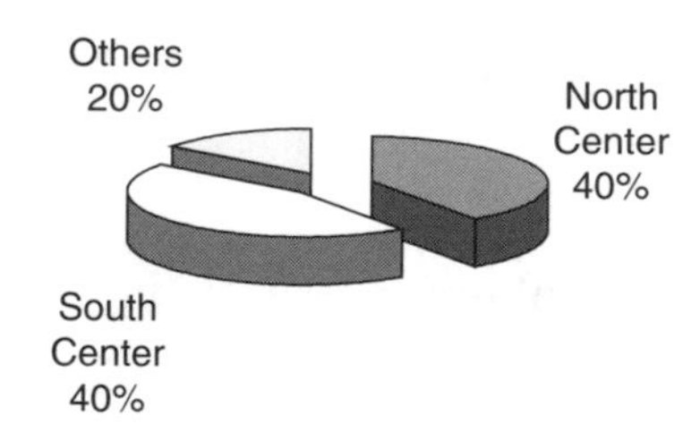

Figure 1.1 Geographic distribution

1.6.2 Human Resources Distribution

The research personnel undertaking nanomaterials and nanotechnology research are mainly located in universities and the Chinese Academy of Science (CAS), which account for over 90% of the overall research potential. There are also research personnel in industries pursuing nanotechnology development, but they account for only 5% and they focus on applications and processing of nanomaterials (Figure 1.2).

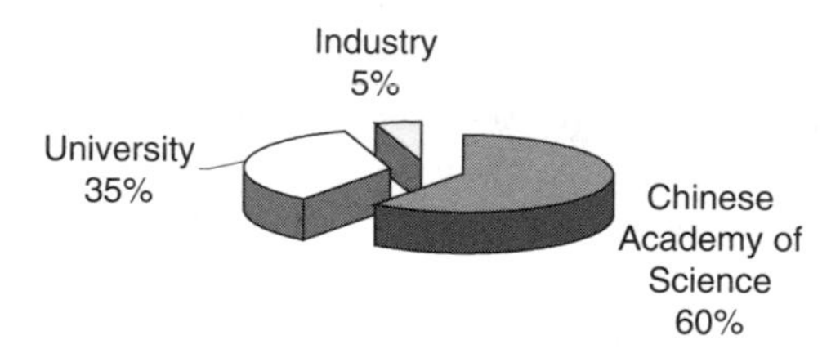

Figure 1.2 Organizational distribution

1.6.3 Personnel Structure

There are more than 4500 scientists in China undertaking R&D of nanomaterials and nanotechnology. Among those research workers, there are about 500 older scientists, 1800 middle-aged scientists and 2200 young scientists. They usually have good qualifications; more than 30% have a PhD technical position or higher, over 40% have a master's degree or mid-level technical position (Figures 1.3 and 1.4).

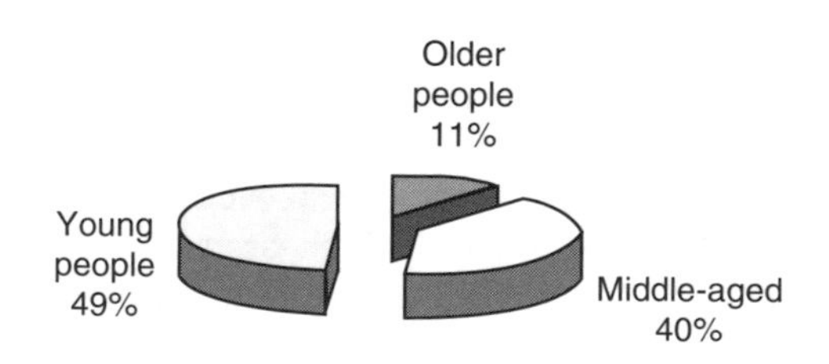

Figure 1.3 Age distribution

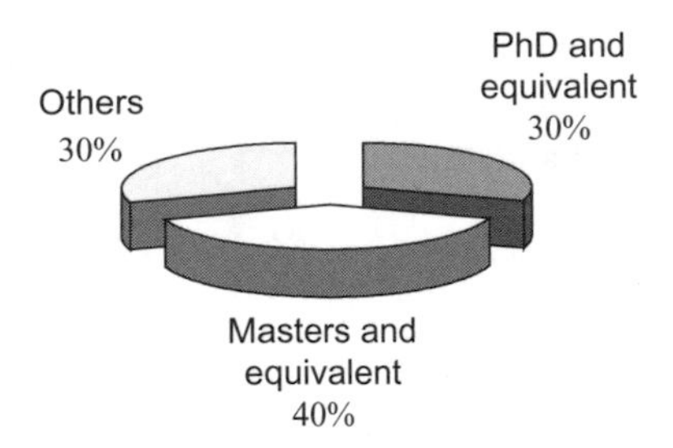

Figure 1.4 Qualifications

1.7 Important Groups and Main Achievements

1.7.1 Research Fields

The main research and development areas in nanoscience and nanotechnology in China are materials, chemistry, physics, information technology and life science. Nanomaterials is one of the most prominentl areas, representing over 50% of all present R&D efforts (Figure 1.5).

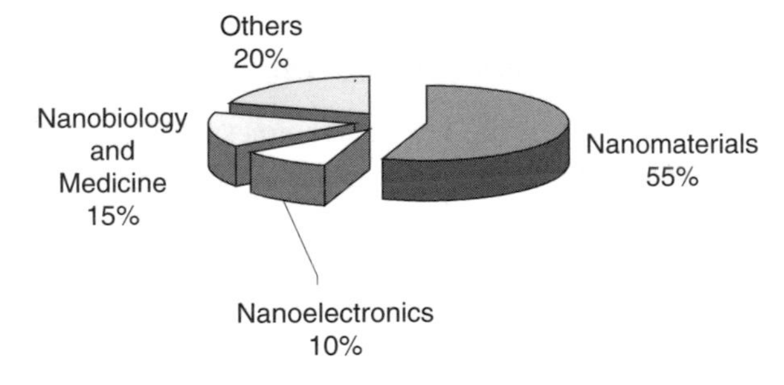

Figure 1.5 Research fields and distribution

1.7.2 Key Achievements in Nanotechnology

Most of the key achievements in nanoscience in China have been reported by the Chinese Academy of Science (CAS) and universities. Private companies plus CAS and university spin-outs are mainly engaged in applications of nanotechnology (e.g. optimizing surface characteristics) and the processing and manufacturing of nanomaterials. Hence most original R&D in nanotechnology in China is still done by CAS and key universities.

The most prominent achievements in Nanotechnology in China are

- oriented synthesis of large-area nanotube arrays;
- synthezising nano nitrogenized gallium using a benzene solvent;
- nanotube arrays on silicon substrates;
- one-dimensional nanowires and nanocables;

- nanodiamond powder using catalytic thermal decomposition;
- first discovery of a rich copper phase;
- functionalized organic nanomaterials.

1.7.3 Key Institutions in Nanotechnology

- Institute of Physics of CAS
- Institute of Solid Physics of CAS
- Shenyang Institute of Metal Research of CAS
- Institute of Chemistry of CAS
- Institute of Technological Physics of CAS
- Shanghai Institute of Silicate of CAS
- Benijing University
- Tsinghua University
- Fudan University
- Shanghai Jiaotong University
- Huadong Technology University
- Huadong Normal University
- Nanjing University
- Sichuan University

1.7.4 Sources of Nanotechnology Funding

Sources of nanotechnology funding are programmes 863 and 973, the National Technology Gong Guan Program and the Natural Science Foundation. There is some funding by industry as applied R&D.

A survey on recent nanotechnology funding (Figure 1.6) indicates that public and private funding for nanotechnology has increased steadily over past years, most of it

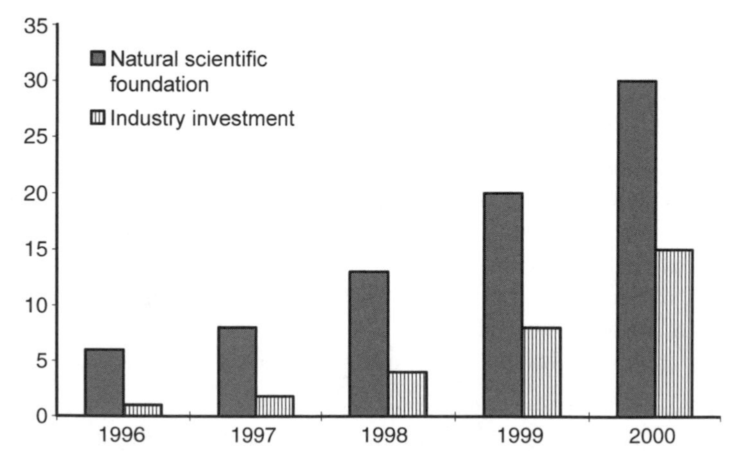

Figure 1.6 Normalized natural scientific foundation and industry investment: base reference of 1 unit is the industry investment in 1996

coming from the public sector. The survey also indicates that industry is paying increasingly more attention to nanotechnology. In 1996 industry investment in nanotechnology was only 15% of National Natural Science Foundation of China (NNSFC) investment but by 2000 it had increased to 50%. One can expect industry investment to outpace government investment some time before 2010.

During 1999 and 2000 there were at least 536 applications to government agencies for nanotechnology funding across six broad disciplines (Figure 1.7). A total of 80 million yuan was allocated to successful grant applications. About 50% of the applications were nanomaterials related, which reflects the general focus on nanomaterials at CAS and Chinese universities. Figure 1.8 shows the number of nanotechnology projects that actually received funding.

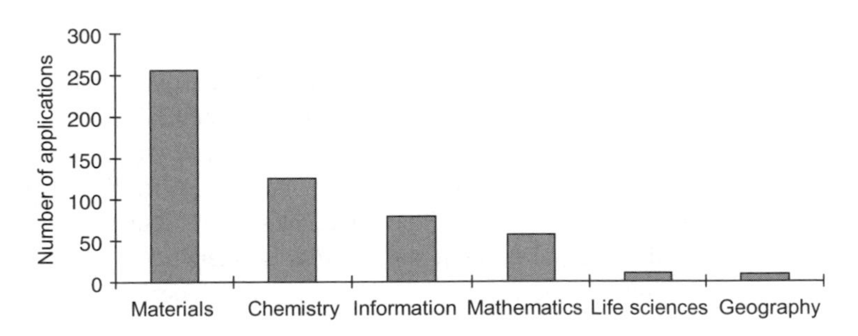

Figure 1.7 Nanotechnology funding applications over the period 1999–2000

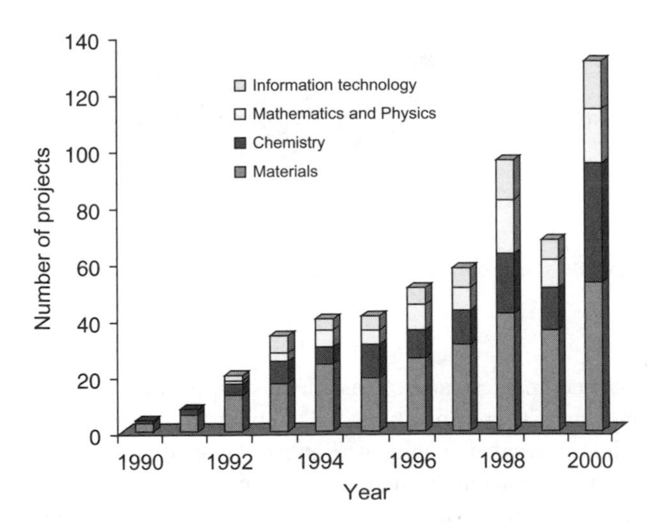

Figure 1.8 Number of nanotechnology projects actually funded

1.7.5 Related Governmental Organizations

To promote development of nanotechnology, national and local government set up corresponding organizations: the Guidance and Harmonization Committee of National Nanotechnology, the National Industrialization Base of Nanotechnology in Tianjing, the Shanghai Industrialization Base of Nanotechnology, the Jiangsu Application and Engineering Center for Nanomaterials, the Shandong Engineering Center of Nanotechnology, the National Industrialization Base of Biological and Medical Nanomaterials in Sichuan, the Jiangsu Engineering Center of Nanotechnology and the Shenyang Industry Park of Nanotechnology.

1.8 Status of Private Nanotechnology Companies in China

1.8.1 Geographic Distribution

Since 1995 some enterprisers have started to go into nanomaterials and nanotechnology. Up to May 2001 there were 323 private nanotechnology corporations in China, 3 billion yuan was devoted to them and three industry areas of nanomaterials and nanotechnology were formed based on Beijing (including Beijing, Tianjing and the north-east), Shanghai (including Shanghai, Zhejiang, Shandong, Jiangsu and Anhui) and Shenzhen (including Shenzhen, Guangzhou and Fujian). Figure 1.9 shows the distribution of private nanomaterials companies in China.

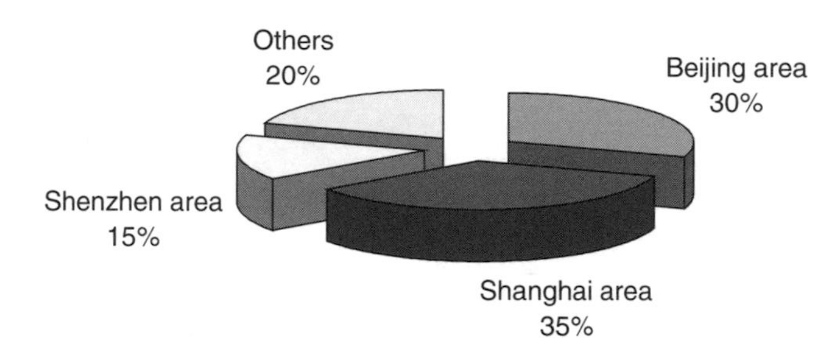

Figure 1.9 Distribution of private nanomaterials companies in China

Nanomaterials companies can be classified as application companies and manufacturing companies. There are about 200 application companies, 95% of them located in Beijing, Shanghai, Zhejiang, Jiangsu, Guangdong, Shandong and Anhui (Figure 1.10). There are about 30 manufacturing companies, about 15% of all nanotechnology companies in China, mainly locating in Shanghai, Zhejiang, Jiangsu, Guangdong and Shandong (Figure 1.11).

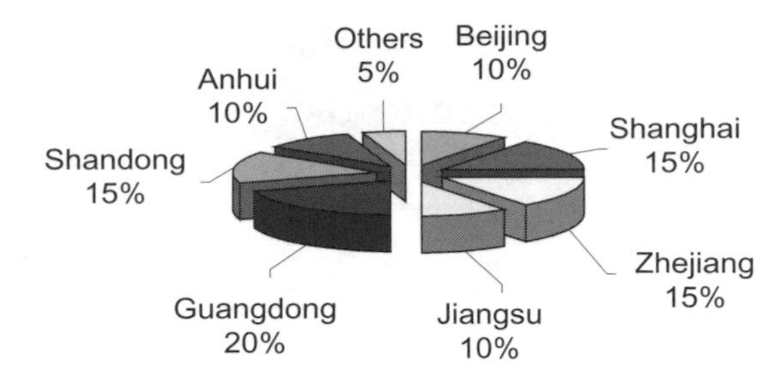

Figure 1.10 Geographic distribution of nanomaterials companies

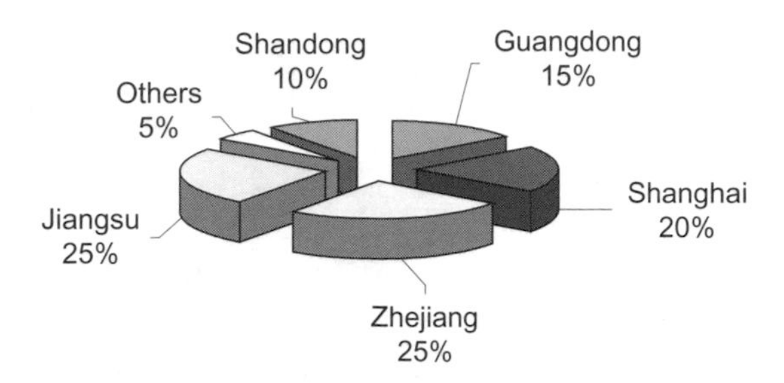

Figure 1.11 Geographic distribution of nanomaterials manufacturing companies

1.8.2 Statistics on Nanotechnology Companies

Figures 1.12 to 1.15 show the statistics (233 companies) on the foundation, ownership and staff numbers of Chinese nanomaterials companies; Figure 1.16 shows the statistics on total assets. At present there are 233 nanomaterials and nanotechnology companies, about half of them were founded after 1995 (Figure 1.12). The distribution of these companies is shown in Figure 1.13 (based on the characteristics), Figure 1.14 (based on the population) and Figure 1.15 (based on the research staff).

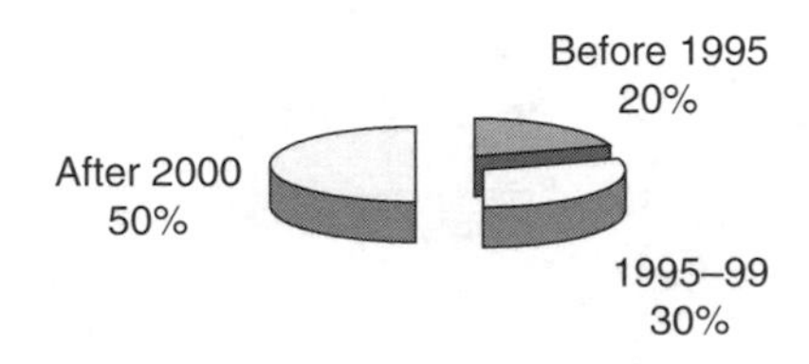

Figure 1.12 Date of foundation for nanomaterials companies in China

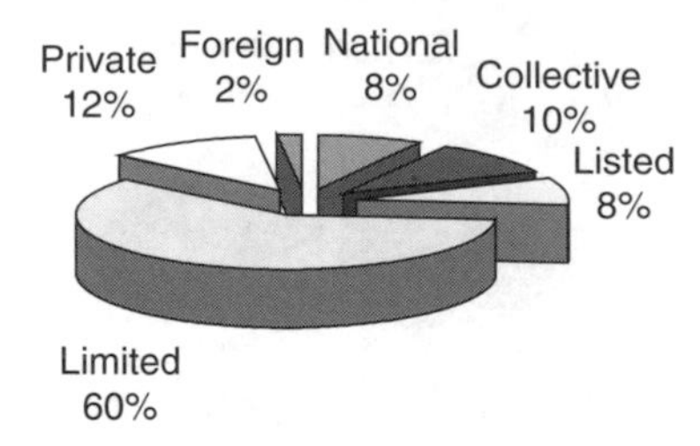

Figure 1.13 Ownership of nanomaterials companies

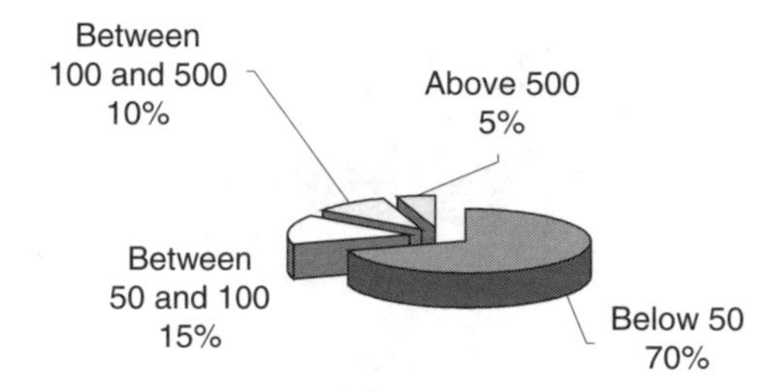

Figure 1.14 Staff numbers of nanomaterials companies

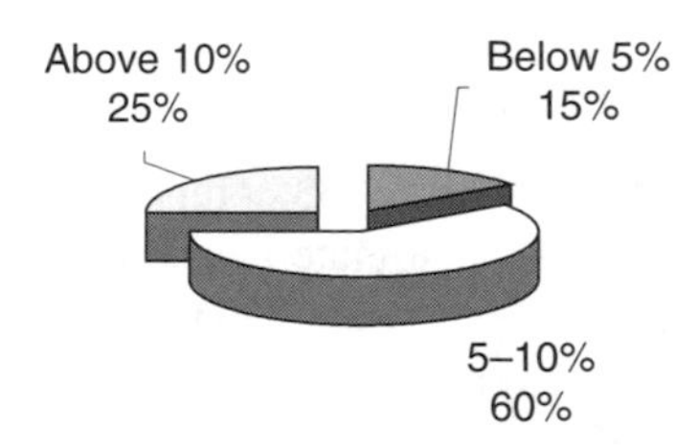

Figure 1.15 Research staff numbers of nanomaterials companies

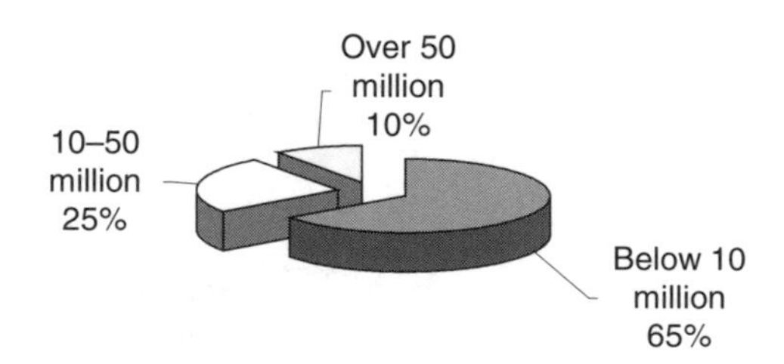

Figure 1.16 Total assets of nanotechnology companies

1.9 Industry Focus and Product Variety

1.9.1 Industry Focus and Product Maturity

Efforts to develop nanomaterials, nanoelectronics and nanomedicine vary across basic research, technology development, industrial manufacturing and market exploitation. The difference in effort is most prominent between basic research and market exploitation (Table 1.1). It is planned to narrow the gap between these two extremes.

Table 1.1 Maturity of key nanotechnology

	Basic study	Technology development	Industrial study	Market exploitation
Nanomaterials	○	○	○	○
Nanoelectronics	○	○	none	none
Nanomedicine	○	○	○	○

Nanotechnology product maturity has been categorized into products which are in pilot testing, batch production and bulk production. Figure 1.17 clearly shows a typical pattern of an emerging technology at an early stage. It is planned to even out this pattern in the near future.

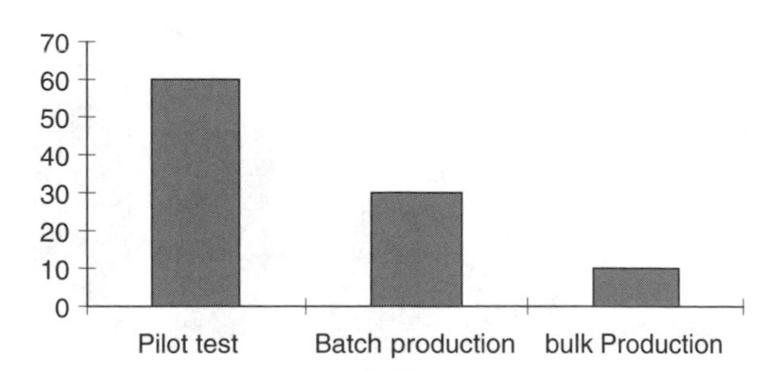

Figure 1.17 Maturity of nanoproducts

1.9.2 Variety and Applications of Nanoproducts

At present there are more than 30 product lines of nanomaterials in China, mostly nano-oxides, nanometal powders and nanocomplex powders. Nanomaterials are mainly designed for use in textiles, plastics, porcelains, lubricants and rubbers (Figure 1.18).

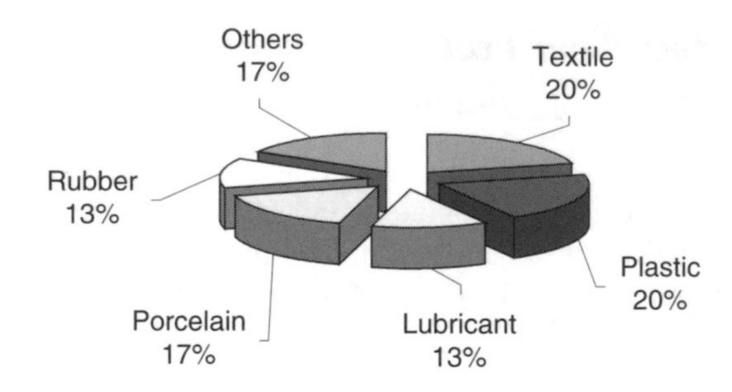

Figure 1.18 Current application areas for nanomaterials

1.10 Funding and Profit Output

Figure 1.19 shows funding and profit output for the whole country. It indicates that the introduced funding gradually increases in recent years, but the profit output increases more slowly than the introduced funding. Furthermore, it is possible that some output values are overestimated as the predicted output value has a direct relationship with the funding amount. A survey of 69 Chinese companies engaged in nanotechnology development showed that the majority of companies (60.8) have a capitalization of less than 30 million yuan. The survey included 13 companies from the Jinghu area, 5 from the South China area, 23 from the Huadong area, 10 from the Huabei area, and 14 from the Middle West area (Table 1.2).

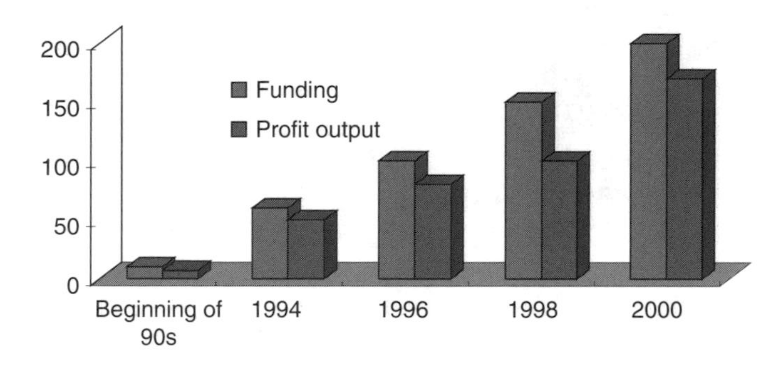

Figure 1.19 Funding input and profit output (million yuan)

Most of the companies with over 100 million yuan capital are those that use nanotechnology to reconstruct their traditional industries, for example, Zhuzhou Horniness Alloy Company (founded in 1954, total capital 1.086 billion yuan, net capital 0.408 billion yuan) and Jiangsu Shenji Corporation (founded in 1979, total capital 0.341 billion yuan, net capital 0.131 billion yuan). There are also some

Table 1.2 Capitalization of nanotechnology companies

Capital (10^6 yuan)	Jingho		South		Huadong		Dongbei		Huabei		Midwest		Totals			
	T	N	T	N	T	N	T	N	T	N	T	N	TC	%	NC	%
>100	0		0		3	1	1	1	1	1	1	1	6	8.7	4	6.8
50–100	2	1	1	1	2	1	0		3	2	1	1	9	13.0	6	10.2
30–50	2		0		7	3	0		2	1	1	1	12	17.4	5	8.5
10–30	2	3	3	2	4	5	1		0	1	3	2	13	18.8	13	22.0
5–10	2		0		3	3	2	2	1	3	6	6	14	20.3	14	23.7
<5	5	6	1	2	4	5	0		3	2	2	2	15	21.7	17	28.8
Total	**13**	10	**5**	5	**23**	18	**4**	3	**10**	10	**14**	13	**69**	100	59	100

Notes: T = number of companies (total capital), N = number of commpanies (net capital), TC = total capital, NC = net capital

nanotechnology companies founded in recent years, e.g. Nei Meng Gu Meng Xi Advanced New Materials (listed company, founded in 1999, total capital 0.606 billion yuan, net capital 0.216 billion yuan) and Hei Long Jiang Zhong Chao Nanotechnology (listed company, total capital 0.104 billion yuan).

There are quite a few small companies with less than 5 million yuan capital, which take up 20% of all companies. Most of them were founded over the past two years. They either develop a technology or a product in collaboration with research institutes. Examples of such collaborations are the Shanghai Aijian Nanotechnology Development Corporation (founded in 2000, total capital 1.94 million yuan, net capital is 1.11 million yuan), which collaborates with the Atomic Energy Institute of the Chinese Academy of Science; the Beijing Eryuan Century Technology Corporation (founded in 2001, total capital 2.1 million yuan, net capital 2.1 million yuan), which collaborates with the Institute of Chemistry; and the Changsha Zhongda Tena Technology Corporation (founded in 2001, total capital 0.483 million yuan, net capital 0.473 million yuan), which collaborates with the Powder Metallurgy Institute of Zhongnan University. Other small companies received funding support through the Innovation Funding scheme such as the Beijing Luborui Lubrication Technology Corporation (total capital 1 million yuan), which received 0.55 million yuan through the fourth Zhongbo Industry Technology Innovation Fund in 2000.

Most surveyed nanotechnology companies (two-thirds out of 69) preferred not to report data on their total and net capital and/or their previous revenue and net profit. However, when asked about future net income expectations, many companies reported an optimistic outlook. Although some of the data in Table 1.3 is still overestimated, these figures at least reflect a general trend in growth of nanotechnology companies.

From Table 1.3 it can be concluded that during 2000 and 2001 most nanotechnology companies gained less than 50 million yuan, i.e. 51.85% of the total number of companies in 1999 and 42.22% in 2000. Some companies reported very small

Table 1.3 Financial performance of nanotechnology companies

Capital (million yuan)	1999		2000		2001	
	Number of companies (income)	Number of companies (net profit)	Number of companies (income)	Number of companies (net profit)	Number of companies (income)	Number of companies (net profit)
Over 100	4		5		4	
50–100	2		3		4	
30–50			1	2	6	2
10–30	5	4	11	2	10	4
5–10	2	1	6		5	4
1–5	8	4	12	11	13	18
Less than 1	6	10	6	13		14
Debt		5	1	5		
Total	**27**	**21**	**45**	**33**	**42**	**42**

profits of less than 1 million yuan, i.e. 47.62% in 1999, and 39.39% in 2000. In 1999 23.81% of the surveyed companies reported debts, and in 2000 only 15.15% reported a negative profit.

Most of the companies with over 10 million yuan profit in 1999 and 2000 are working in traditional industries, such as Taian Kangping Floss Textile Corporation, Jiangsu Sujing Corporation, Jiangsu Changtai Chemical Engineering Corporation, Xiamen Tungsten (listed company) and Zhuzhou Horniness Alloy Company. The reported profits did not, however, come from the introduction of nanotechnology products. Two corporations which produce calcium carbonate nanopowder (Guangping Chemical Engineering Corporation and Nei Meng Gu Meng Xi Advanced New Material Corporation) used this technology to reconstruct their traditional business into a nanomaterials-focused business.

Data: Ministry of Science & Technology, P.R. China

2

Current Status of Nanotechnology in Korea and Research into Carbon Nanotubes

Jo-Won Lee[1] and Wonbong Choi[2]
[1]*Korean National Program for Tera-level Nanodevices and*
[2]*Florida International University*

2.1 Introduction

Despite the recent economic uncertainty, enthusiasm to develop high-tech industries still runs high across the world. Specifically, many advanced countries are putting aside most of their investment in research projects, since a high value-added technology can only be obtained through time-consuming and costly research. Korea is also following this trend. Fortunately, the Korean government, here after called 'the government', has designated nanotechnology (NT) as one of six important fields that would be the growth engine for the next 10 years. The other five fields are information technology (IT), biotechnology (BT), environmental technology (ET), space technology (ST) and contents technology (CT). Back in July 2001 the government formulated an ambitious ten-year master plan to nurture NT, which is an initial step to keep up with the global trend in favour of the next-generation technology. The first part of this chapter gives a detailed description of the current status of NT in Korea. Among the many activities in Korea, carbon nanotube research has revealed tremendous potential for future electronic device

Nanotechnology: Global Strategies, Industry Trends and Applications Edited by J. Schulte
 ISBN: 0-470-85400-6 (HB)

applications. The second part of this chapter describes research into carbon nanotubes for nanoelectronics.

2.2 Current Status of Nanotechnology in Korea

Korea is renowned for its excellence in some high technologies and large-volume process engineering, which is shown by its world-leading position in semiconductor memory chips, shipbuilding and many electronic products. In addition, Korea is somehow a leader in information technology (IT). At the end of December 2001, the number of mobile phone users (28 million) exceeded that of PC users (17 million) and almost half of all Koreans (24 million) used the internet. This shows that the country is at the forefront of utilizing state-of-the-art technologies.

In July 2001 the government drew up a ten-year plan for nanotechnology. It breaks down into three stages until 2010 whereby the government is going to pour 1.48 trillion won ($\$1 = \sim 1200$ won) into the scheduled projects (Table 2.1). The government's aims are to pave the way for the introduction of NT infrastructure within five years and to secure core NT for entering the world's top five nations in this field by 2010, although Korea's present achievements in NT are very few, at 25% of the rating for advanced countries. However, we believe that the technological expertise accumulated during the past decades in semiconductor devices, processing and manufacturing could provide a launching pad for NT.

The government will focus on the selected areas that have the most commercial potential and competitiveness compared with advanced countries. The promising fields are nanodevices, nanomaterials, nanoprocessing and other basic technologies. The government will execute the plan to obtain at least 10 cutting-edge NTs and to produce 12 600 NT experts by 2010.

NT in Korea is largely in its infancy, hence there is a great shortage of trained engineers. According to a recent survey, Korea has around 1000 NT scientists and engineers. This number emerged suddenly one morning when many pseudo-nanoscientists and engineers claimed their work was NT. Therefore one of the major focuses of the NT plan is to foster as many highly qualified NT scientists and engineers as possible. Consequently, the plan also includes the creation of interdisciplinary programmes devoted to NT by multiple departments at major universities and the re-education for NT fields of researchers in traditional disciplines.

Under the plan, the government is supposed to create a centralized nano fabrication centre where all the research facilities are open to domestic and foreign scientists and engineers from university, industry and national labs on a peer-review basis, while pushing through the establishment of a facility network domestically and with foreign countries.

In 2002 Korea invested 203.1 billion won in NT and introduced a bill that would accelerate NT developments. The move is a reflection of the government's view that NT will be one of the most important fields for Korea in coming years. The 2002 investment figure of 203.1 billion won is a 93.1% increase from 105.2 billion won

Table 2.1 Ten-year nanotechnology investment plan in units of billion won

Classification	First phase (01 to 04)			Second phase (05 to 07)			Third phase (08 to 10)			Total
	Government	Private	Subtotal	Government	Private	Subtotal	Government	Private	Subtotal	
Research	233	50.5	283.5	267	158	425	267	237	504	1212.5
Manpower	35.5	—	35.5	26.5	—	26.5	21.5	—	21.5	83.5
Facilities	73.6	31.8	105.4	32.7	12.6	45.3	26.7	11.6	38.3	189
Total	**342.1**	**82.3**	**424.4**	**326.2**	**170.6**	**496.8**	**315.2**	**248.6**	**563.8**	**1485**

in 2001. Of the total, the government has set aside 160.1 billion won for research and development, 34.6 billion won for a centralized nano fab build-up and 8.4 billion won for engineer education programmmes. The government will also seek NT industrialization support funds, about 3 billion won for the planned construction of the nano fab. Note that the budget falls far short of those in the US and Japan, although the government plans to invest heavily in NT.

Korea's research into NT has yielded fruitful results for the past decade. According to a recent report from Thomson ISI, 579 papers on NT by Korean scientists from 1991 to 2000 have been published in academic journals across the world. Some of them have been printed in the world's top scientific journals, such as *Science* and *Nature*. Most of the accomplishments were achieved in nanoelectronics, nanoprocessing and nanomaterials, whereas advances in nanobiological research are still disappointing.

Probably more than 100 big companies and ventures in Korea are engaged in NT but this number will increase as time goes on. Big companies are concentrating on improving their core products (IT areas) and creating new business from NT, while most ventures are relatively recent and specialize in nanomaterials. Except for nanobiology, Korea is rather trailing the worldwide trend, which focuses on research and development of nanoelectronics and nanomaterials.

By looking into the patent rate of NT in Korea, we can evaluate the current NT research capacity. Sometimes the patent rate can be a more reliable indicator than the number of published papers, since patents have to go through a longer approvals process before they are published. It was found that a total of 542 patents had been granted to NT applications from 1997 to 2001. The number is rather meagre during 1997 to 1998, but has been exponentially increasing at a rate of 54.3% since 1999. Most of the patents are related to carbon nanotubes (CNTs) and their applications filed by companies. This indicates that companies play a major role in NT industrialization whereas universities and national labs are the mainstay that drive Korea's research in this area. This is partly due to the fact that most early NT funding has long been awarded in this area, helping to create a series of outstanding achievements by Korean scientists.

Almost all aspects of research related to NT are carried out by various groups in national labs, universities and industries covering nanodevices, nanomaterials nanobiology and basic technologies. Although some have made outstanding achievements in labs, there is already some confusing information about results from NT. Some pseudo-NT scientists and engineers misled the public just by adding 'nano' to their work or their products.

A national programme named Tera-Level Nanodevices (TND) was established in April 2000. Its visions are to strengthen the national competitiveness in nanoelectronics and to overcome the technological limits imposed on upcoming semiconductor technologies. The TND programme is one of the government's key NT programmes born out of Korea's 21st Century Frontier R&D Program and funded by the Ministry of Science and Technology (MOST). The TND is a ten-year programme consisting of three phases. The first phase will be operated as a versatile basic cell development for tera-level nanodevices. In the second phase, major efforts will be made towards an integration process for nanoscale devices. The third phase will concentrate on developing tera-level integrated arrays of nanodevices.

The TND has a total strength of 180 PhD, 120 MS and 200 graduate students from leading universities, national labs and industries. They are from physics, materials science, chemistry and engineering. The total budget was about 17 billion won for fiscal year 2002. This budget increases gradually each year. Actual R&D is subcontracted through the TND to universities, national labs and industries. The TND covers four major areas: tera-level nanoelectronics, spintronics, molecular electronics and core technologies (Table 2.2). In addition, the feasibility studies are undertaken for high-risk subjects in the nanodevice field.

Table 2.2 Projects operated by TND

Tera-level nanoelectronics
Tera-bit-level single-electron memory
Nano CMOS
SET logic and RF SET
Terahertz-level IC
Spintronics
MRAM integration process
MR material and single-cell process for MRAM
Spin injection devices
Molecular electronics
Terabit-level carbon nanotube devices
Terabitqevel organic devices
Core technologies
Nanopatterning
Nanodeposition
Nanoanalysis
Tera-level optical interconnection

In addition to the TND, the government has initiated two major NT programmes in 2000 as part of the 21st Century Frontier Program. A total of 200 billion won will be invested during the next 10 years. The objectives are to develop seed technologies that will produce the functional nanomaterials and the nanomechatronics for producing 10 nm level nanoprocessing.

The government has drafted a strategic plan for R&D and infrastructure build-up to expedite NT commercialization. The R&D programme consists of core technology and base technology. Core technology is five projects, including tera-level storage, and will receive more than 1.5 billion won per project for six years. Meantime, nine projects including nanobiochip, will be conducted for the base technology, with less than 1 billion won per project for five years. The R&D funding emphasizes interdisciplinary research through a mandatory collaboration between different disciplines from all sectors of the research community.

As part of the plan to establish the infrastructure, the Korea Advanced Institute of Science and Technology (KAIST) was selected by the government in July 2002 to build the 200 billion won large-scale nanofab centre. The main focus of the centre will be to foster NT experts and offer NT-related services and research equipment. It is due to be completed in 2005, and then Korea will be able to carry out world-class NT research.

Several other NT-related national programmes are now running. For example, in 2001 MOST set aside 14 billion won for 12 creative research centres to produce world leaders in NT fields and for 7 science or engineering research centres; they existed only in the university to promote collective works. Some 38 national research labs have also been established, with a total of 9 billion won for fiscal year 2002.

In academia, Seoul National University (SNU), KAIST, Hanyang University and others are conducting fundamental research to understand the behaviour of nanomaterials, nanoprocessing and nanodevices. A basic understanding of their behaviours can lead to new devices and new nanostructures. For example, Professor Young Kuk at SNU and his colleagues reported in *Nature* a method for inserting carbon fullerene structures into a nanotube, breaking it up into multiple quantum dots with lengths of 10 nm (Figure 2.1). The technique could be used to construct nanoscale ICs and optoelectronic devices [1]. Another SNU professor, Taeghwan Hyeon, and his coworkers demonstrated uniformly sized iron nanoparticles (4–16 nm) using a new synthesis (Figure 2.2). The method is recognized by

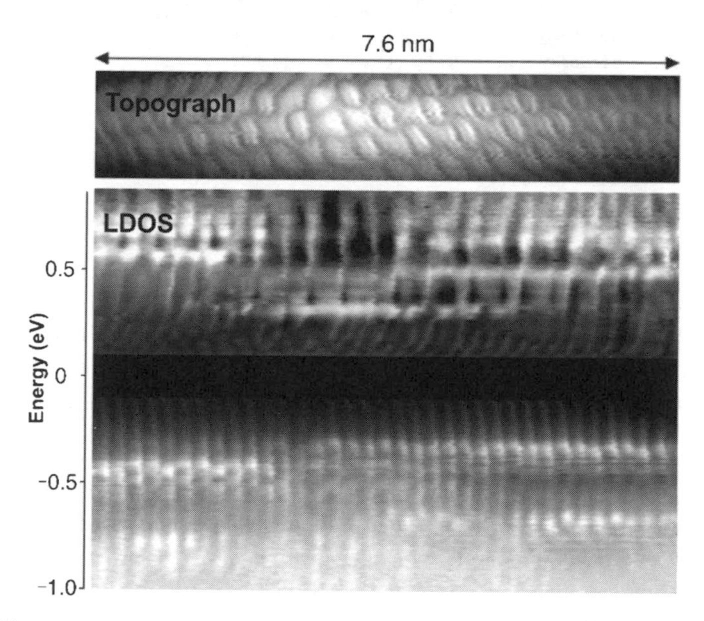

Figure 2.1 Atomically-resolved scanning tunnelling spectroscopy showing the local density of states around a semiconducting carbon nanotube intramolecular junction. Different band gaps and a localized defect state are observed revealing their spatial variation

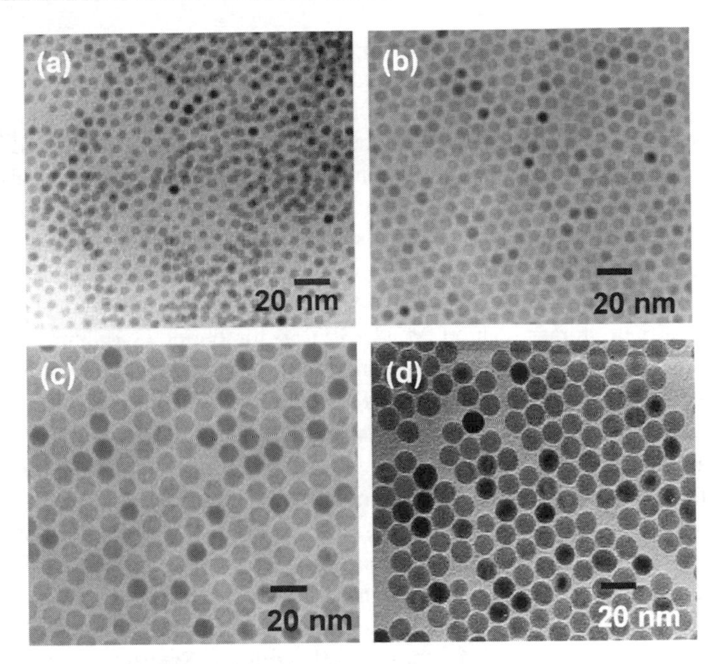

Figure 2.2 Transmission electron microscopy (TEM) images of monodispersed iron oxide nanocrystals: particle size (a) 4 nm, (b) 7 nm, (c) 11 nm, (d) 13 nm

many researchers in the world to be adopted as a new standard for the preparation of Fe_2O_3 nanoparticles [2]. In addition, Professor Hai-Won Lee and his colleagues at Hanyang University revealed a new method to increase the speed of atomic forcemicroscope (AFM) lithography using their own resist. The patterning speed is 2 mm/s, 100 times faster than others. This is a promising result, leading to the possibility of using AFM lithography on larger wafers (Figure 2.3) [3]. Several

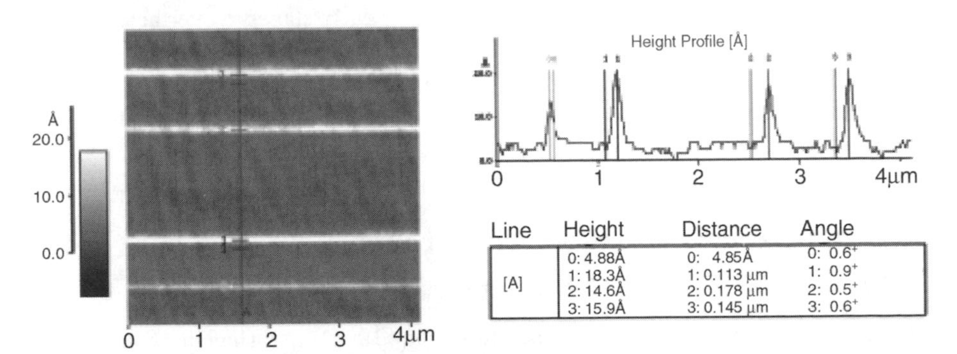

Figure 2.3 Topographic image of a line pattern on a silicon wafer using the mixed self-assembled monolayer (SAM) resist (DAD·2HCl and TDA·HCl) at the high lithographic speed of 0.5 mm/sec

leading universities have implemented interdisciplinary programmes associated with NT for MS and PhD students and even allow NT departments to attract undergraduate students.

In preparation for future electronics, the Korea Institute of Science and Technology (KIST), the country's premier national lab, located in Seoul, is concentrating research on nanomaterials, nanophotonics, NEMS, MRAM and spintronics using spins and electrons. Nanomaterials and spintronics will be its main focus for the next 10 years. The Electronics and Telecommunications Research Institute (ETRI), most famous for the world's first CDMA development, is now focusing on ultra high density data storage, nano-CMOS, SET, semiconductor quantum structures and new functional quantum devices. Many other national labs, including the Korea Institute of Machinery and Materials (KIMM), the Korea Research Institute of Standards and Science (KRISS), the Korea Electronics Technology Institute (KETI), and so on, are also increasing their research activities in NT fields such as nanomaterials, nanoprocessing, instruments and energy-related technology.

In the industrial sector, a research team at the Samsung Advanced Institute of Technology (SAIT) unveiled the world's first 4.5 in field emission display (FED) using single-walled carbon nanotubes in 1999 [16]. Cooperating with Samsung SDI, in 2002 it made a significant improvement with a full-colour, wide, VGA-type, 32 in FED that can produce a brightness of 200 cd/mm^2. Samsung's researchers are also exploring tera-level SET memory, MRAM and CNT transistors. In 2001 SAIT demonstrated the world's first vertical CNT field-effect transistor (FET). This is the only one fabricated using a top-down approach instead of bottom-up. Early in 2002 it demonstrated non-volatile memory operation based on the CNT-FET. The LG Electronics Institute of Technology is conducting research into photocatalysis and CNT FEDs. It is also exploring an ultra high density data storage system based on scanning probe microscopy, which may enable data densities well beyond the current storage density of magnetic recording. Several other big companies, such as Samsung Electronics, Hynix, SK, Hyundai Motors and Iljin, are also involved in NT research to improve their products and create new business.

Commercial applications of NT are still in their early stages. Nevertheless, there is little doubt that NT is expected to bring revolutionary breakthroughs for almost all technologies. It is also expected to create exceptional earnings potential and new business opportunities in electronic materials, communication, environment, energy, medicine, and so on.

2.3 Carbon Nanotube Research in Korea

2.3.1 Background of CNT Research

Since the discovery of carbon nanotubes in 1991 by using high-resolution transmission electron microscopy (HRTEM), there have been intensive research activities in the area of carbon nanotubes (CNTs), not only because of their fascinating properties, but also because of their potential technological applications.

Nanotubes show exceptional electronic and mechanical properties together plus nanosize diameter and hollowness. They behave like one-dimensional quantum wires that can be either metallic or semiconducting, depending on their chirality and diameter. High current-carrying capacity and heat dissipation together with structural robustness are attractive properties for future nanoelectronics. There is increasing interest in applying carbon nanotubes for nanoelectronics, FEDs, hydrogen storage, fuel cells, supercapacitors and gas sensors [4, 5]. Needless to say, the realization of nanotubes for use in everyday life depends on turning them into devices.

To increase their speed and memory capacity, silicon transistors have been developed by downscaling the device dimensions and increasing the charge concentration. These two changes have been a major focus of device development for the past 10 years.

Figure 2.4 shows a possible path for further shrinkage in DRAM technology. However, this continuing shrinkage causes several serious problems. In particular,

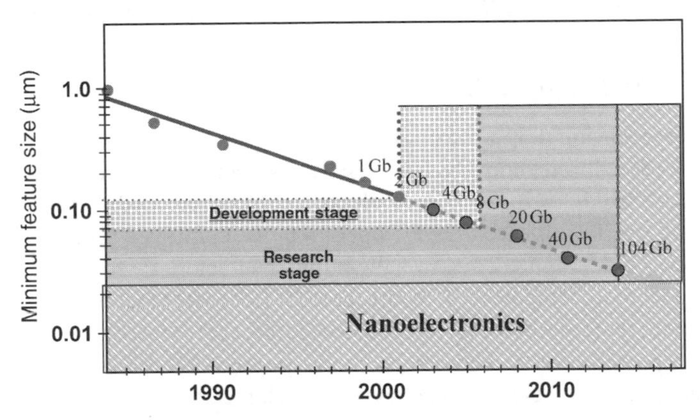

Figure 2.4 The minimum feature size of CMOS plotted against year, which is modified from the 2001 International Technology Roadmap for Semiconductors

the small amount of free charge to be detected has been a major focus of new device development for the past 10 years. Some limitations of shrinkage are (i) high electric field breakdown due to a bias voltage being applied over very short distances, (ii) malfunctioning due to the limit of heat dissipation for any type of densely packed nanodevices, (iii) overlapping of the depletion region, which results in quantum mechanical tunnelling of electrons when the device is turned off, (iv) non-uniformity of doping on small scales and (v) shrinkage and unevenness of the gate oxide layer causing leakage current from gate to drain. In order for a FET device to operate on the nanometre scale, it is desirable to have a device that does not depend on the doped materials and that operates on a quantum mechanical basis. Carbon nanotube appears to be a candidate for overcoming the limitations of downscaling.

It has been reported that using a CNT as a FET channel can change the conductivity by a factor of 1000 or more. It is also expected that CNTs could solve the thermal dissipation problem due to their high thermal conductivity. In addition, the transconductance of a CNT-FET has been reported as more than four times higher than for a silicon MOSFET. CNTs are expected to have ballistic transport, which means no scattering occurs during charge transport [6–8]. They can transport terrific amounts of electric current without the doping problem of silicon FETs, because the bond strength between carbon atoms is much stronger than in any metal. It was reported that multiwall carbon nanotubes (MWNTs) could pass a very high current density up to 10^{10} A/cm^2. Several papers have recently reported on CNTs for FETs, CNT-logics and memory operation. However, most of the results are based on one or several units of CNT-FETs. There are still many obstacles to device realization, such as aligning CNTs, controlling the electron energy band gap of CNTs, integration, and reliability. Our research is focusing not only on device realization but also on developing technology for CNT functionalization. The vision of this project is to make future electronic devices entirely out of CNT devices such as CNT transistors, CNT memory, and interconnects.

2.3.2 CNT Field-Effect Transistor

Since the first working device was reported in 1998, the number of papers on CNT-FETs has increased tremendously. CNT-FETs have been made either by employing a back gate electrode or by a top gate electrode on top of a silicon wafer covered with an insulator. To improve the FET operation, we employed a top gate structure with thin gate oxide. Figure 2.5 shows the output characteristic for a CNT-FET with top gate and an oxide thickness of 28 nm. The CNT is passivated by an oxide film so the atmosphere does not influence the electrical transport property of the CNT, as in previously reported results. The device shows p-type CNT-FET behaviour, where current increases with increasing negative gate voltage and decreases down to a few femtoamperes (fA) with positive gate voltages. The ratio I_{on}/I_{off} is over 10^5 at $V_{sd} = 1$ V while the gate voltage was swept from -4 V to 4 V; in the off state, the current remained less than a few picoamperes. The low off-state current is attributed to the geometry of the top gate electrode and the high quality of the oxide film. The operating temperature of a CNT-FET depends on the energy band gap of CNTs, which is directly related to how the CNTs are made. Higher performance of a CNT-FET is expected by using higher-quality CNTs or by reducing the thickness of the gate oxide.

2.3.3 Selective Growth of CNT

Future integration with conventional microelectronics, as well as development of novel devices, requires that CNTs can be grown in highly ordered arrays or located at a specially defined position, such as predeposited catalyst pads or a partially exposed nanotemplate. To get highly ordered CNTs in the selective area, an anodic

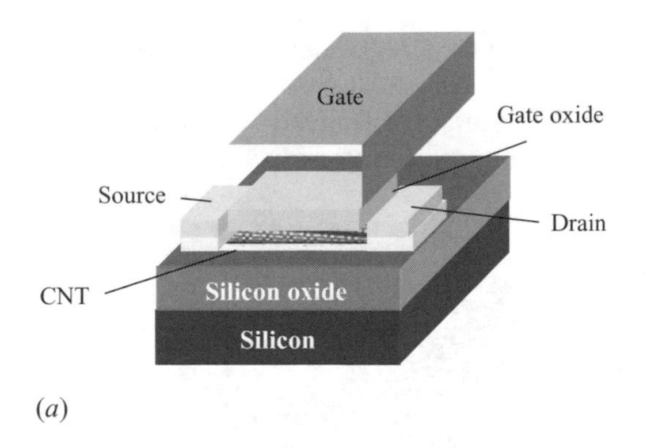

(*a*)

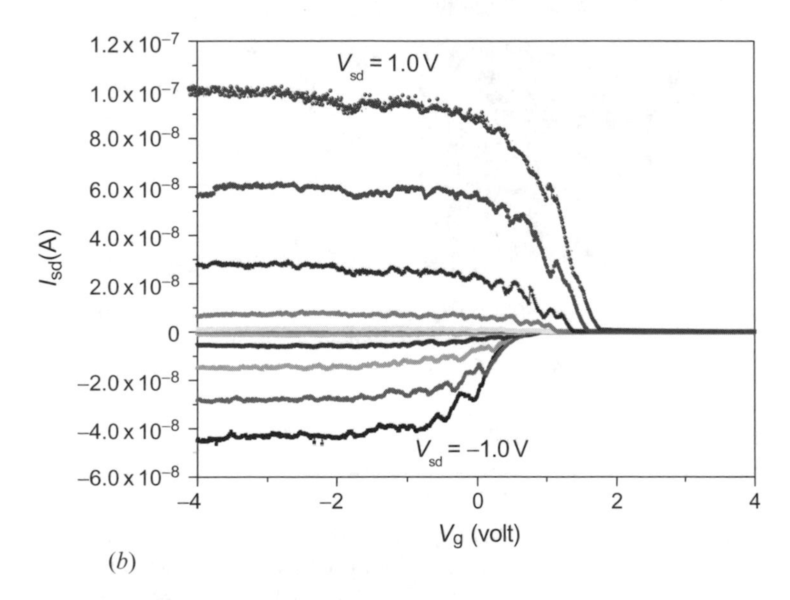

(*b*)

Figure 2.5 (a) Schematic of CNT-based field-effect transistor (FET) with top gate electrode. (b) Drain current versus gate voltage of a CNT-FET with top gate and an ONO layer of 28 nm. The device is a p-type CNT-FET

aluminum oxide (AAO) template was employed and a vertical transistor was fabricated [9, 10].

A vertically aligned transistor is fabricated in the following steps: nanopore formation by anodization, CNT synthesis, metal electrode formation, oxide deposition and patterning, gate electrode formation. A transistor can be constructed as small as the diameter of the CNT. The SiO_2 was deposited at the top of aligned CNTs and followed by e-beam patterning, so that the electrode attached to the CNTs only through the patterned holes. The gate oxide of SiO_2 was deposited followed by deposition of the top gate electrode. The transistor unit cell can be made as small as the diameter of the CNT, which corresponds to the tera-level CNT transistor with a density of 10^{12} elements per square centimetre.

Figure 2.6 shows SEM images of selectively grown carbon nanotube arrays which have been ion milled to remove residual amorphous carbon from the

Figure 2.6 (a) Cross section of AAO template. (b) SEM image of top surface of AAO template. (c) Cross section of three-dimensional nanotube blocks or towers grown selectively. (d) SEM image of vertically aligned CNTs grown in the patterned nanopore

template surface and then partially exposed by etching the alumina matrix using a mixture of phosphoric and chromic acids. Ohmic contact between the CNT and the electrode is required for the CNT-FET to operate. To improve the contact property of the CNT/metal interface, rapid thermal annealing (RTA) was performed. Carbide formation during RTA may enhance the contact property and produce ohmic contact. In the integrated device, each CNT is electrically attached to the bottom electrode (row) and upper electrode (column), and the gate electrode is positioned over the top electrode. Each intersection of two electrodes, bottom and upper, corresponds to a device element with a single vertical CNT. Figure 2.7(a) The gate metal is deposited right after the oxide deposition over the drain electrode. A top view of an $m \times m$ device array is shown in. At each point (n, m) in the array, the vertical CNT is used as a current channel. The speed of the on/off switch depends on the frequency of array sweeping, in which the on state corresponds to where both the intersection point and gate electrodes are turned on.

2.3.4 Bandgap Engineering

It has been reported that semiconducting CNTs show p-type semiconductor. To perform the logic functions, both p-type and n-type CNT-FETs are required.

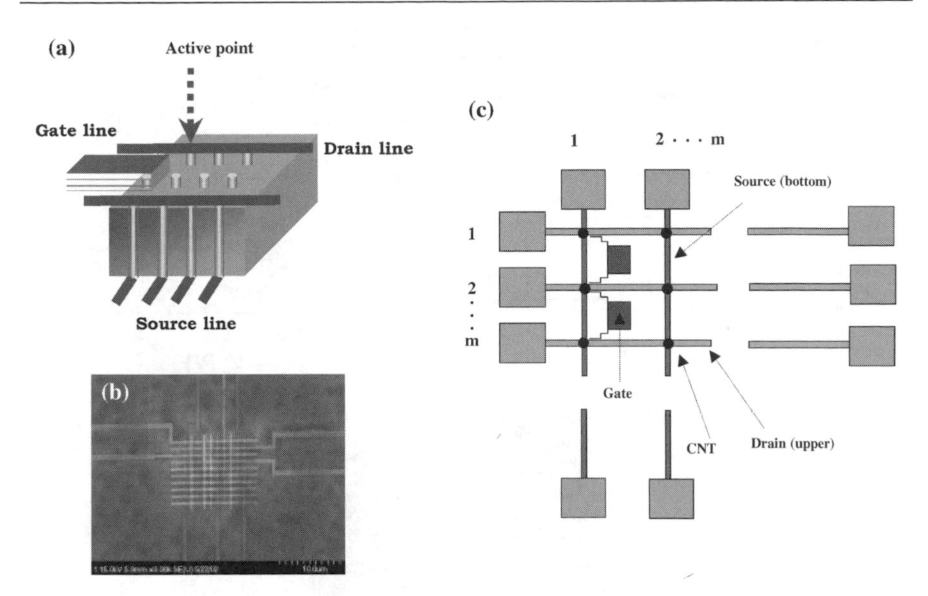

Figure 2.7 (a) Device architecture of a vertical CNT transistor. One device unit consists of a CNT, at the intersection of the top and bottom electrodes. (b, c) Top view of an $m \times m$ fabricated device array and its schematic diagram. CNTs are located at the intersection of a drain and a gate electrode

Recently, it has been discovered that the electrical properties of CNTs can be modified by chemical doping using various molecules. Examples of functional modification are functionalizing CNTs by doping with potassium and annealing in oxygen. We have developed unique technology for hydrogen functionalization of CNTs that leads to the transformation of metallic (narrow-gap semiconducting) CNTs to semiconducting (large-gap semiconducting) CNTs (Figure 2.8) [11]. We demonstrate this phenomenon by fabricating a heterojunction between the pure CNT and the functionalized CNT, which clearly shows rectifying and gating effects from the metallic CNT at room temperature. It was attributed to the C–H bond inducing sp^3 hybridization and thus removing the π and π^* bands near the Fermi level, opening the energy gap.

Logic gates and ring oscillators with n-type and p-type nanotube FETs have been reported [12, 13]. The performance of nanotube logic circuits is still far behind that of silicon-based logic circuits, but it will be improved by enhancing the fabrication processes in the near future.

2.3.5 CNT Memory

CNTs could be used not only as a switching device and interconnect wires, but also as a memory device. This is doen by fabricating a non-volatile memory based on CNT-FETs and oxide–nitride–oxide (ONO) storage nodes. The charges are stored

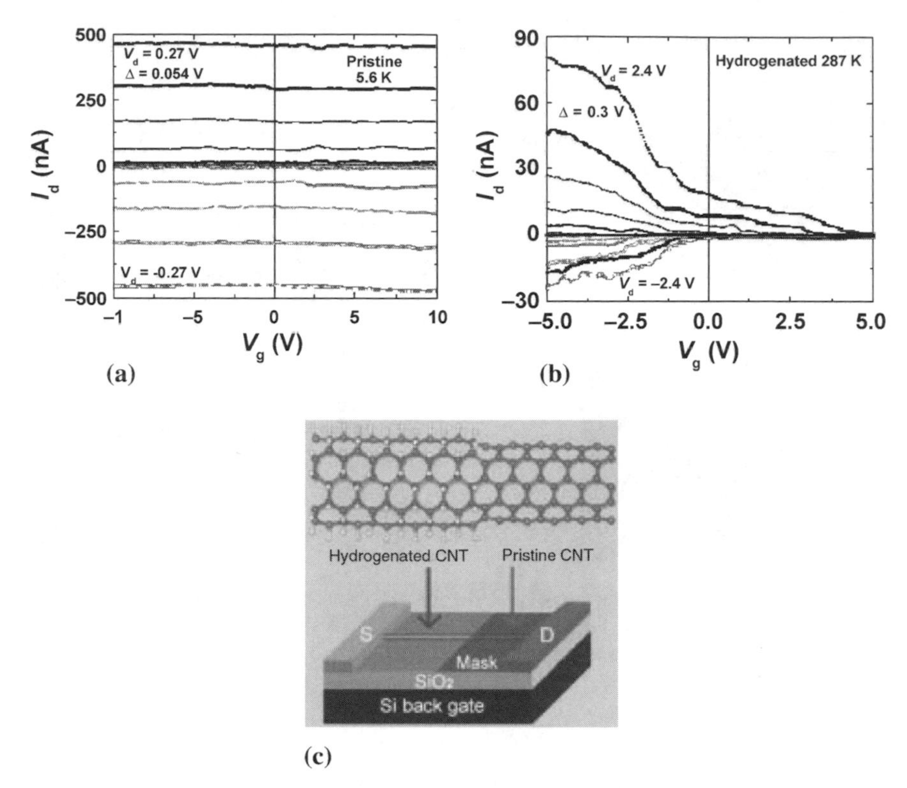

Figure 2.8 Source–drain current as a function of gate voltage at different drain voltages. (a) A pristine metallic/semiconducting (M/S) sample showing no gating effect at 5.6 K; (b) M/S sample after hydrogenation, showing gating and rectifying effects at room temperature; (c) schematic of the CNT–metal contact, where half the CNT was buried by SiO_2 with a thickness of 100 nm

in ONO traps as typically observed in SONOS memory. The stored charges increase the threshold voltage with a quantized increment of 60 mV, suggesting that the ONO has traps with quasi-quantized energy states. The quantized state is related to the localized high electric field associated with a nanoscale CNT channel. These results strongly indicate that the CNT memory can be a candidate for ultra high density flash memory [8].

For flash memory operation, a large threshold voltage shift is essential in obtaining large values of I_{on}/I_{off}. It was reported that reducing the channel width in a MOSFET increases the threshold voltage shift. Therefore CNT-based memory devices are expected to have higher-performance memory operation. Among the memory charge films, an SiO_2–Si_3N_4–SiO_2 (ONO) layer is known to have high breakdown voltage, low defect density, and high charge retention capability [14, 15]. Therefore ONO has been used as the dielectric in dynamic random access memory (DRAM) and electrically erasable programmable read-only memory

(EEPROM) devices. We have presented a novel structure for CNT-based non-volatile memory employing the CNT as a nanometre channel with ONO charge node.

The structure of CNT flash memory is shown in Figure 2.9(a). A charge storage node, consisting of ONO, is located between the CNT and the gate electrode. The memory node is deposited onto the CNT followed by deposition of the top gate electrode. The Si_3N_4 film is known to contain a large number of charge traps, hence it provides a low-potential site for storing charges. The bottom oxide between Si_3N_4 and CNT must be thin so that charges are injected and removed easily through tunnelling. The thick gate oxide of 14 nm was deposited between the nitride film and the gate electrode to suppress charge injection from the gate electrode, so the injected charges from the CNT could be kept at the nitride film.

The measured drain current as the gate voltage was swept up and down revealed clear hysteresis. The threshold voltage shift is about 2 V when the gate sweeping voltage is at 12 V. This suggests it will be possible to create non-volatile memory based on CNT channels. It has been reported that the operating temperature of a CNT-FET depends on the electron energy band gap of CNTs. We have tested several devices operating from low temperature to room temperature, depending on the quality of the CNT. The measured drain current is shown in Figure 2.9(b) as the gate voltage was swept up and down. Obvious hysteresis occurred when the gate voltage was swept over 4 V. The threshold voltage shift is about 2 V when the gate sweeping voltage is at 12 V. The hole density is estimated by calculating the CNT capacitance per unit length with respect to the top gate, $C/L \sim 2\pi\varepsilon\varepsilon_0/\ln(2h/r)$, where h is the thickness of the ONO, L is the length between source and drain electrodes, and r is the radius of the CNT. Taking the effective dielectric constant, ε, for the ONO layer as $\sim$3, $h = 30$ nm, $r = 1.5$ m, $L = 1\mu$m and the depleting gate voltage, V_{gd} as 2 V, we can obtain the hole density as $p = 580\ \mu m^{-1}$. The hole mobility, μ_h, can be calculated by considering the transconductance of the CNT-FET in the linear regime using the relation $\mu_h = \frac{1}{V_{sd}}\left(\frac{dI}{dV_g}\right)\left(\frac{L}{C}\right)L$. The transconductance ($dI/dV_g$) at $V_{sd} = 0.1$ V is $\sim$13.5 nS. So the calculated hole mobility is $\mu_h = 29\ cm^2\ V^{-1}\ S^{-1}$. This value is higher than for the single-walled nanotube (SWNT) and lower than for the multi-walled nanotube (MWNT) reported by Martel *et al.* [6]. The memory operation was characterized by measuring the threshold voltage shift after charging the ONO film; the threshold voltage is defined as the gate voltage at which the current reaches 5 nA) (Figure 2.10). The applied positive gate voltage increases the threshold voltage, indicating that holes are injected from the CNT to the ONO film, so that trap sites are occupied by holes. For 0 to 7 V charging voltage pulses, the shift in threshold voltage was quasi-quantized with an increment of 60 mV. Since the *diameter* of the CNT is about 3-nm, these gate voltages produce a high electric field around the surface of the CNT. Using the image charge method, we calculate the electric field near the CNT as shown in Figure 2.11, where the gate is represented as a perfect conductor and the ONO layer between the CNT and the gate is considered as a single layer with the effective

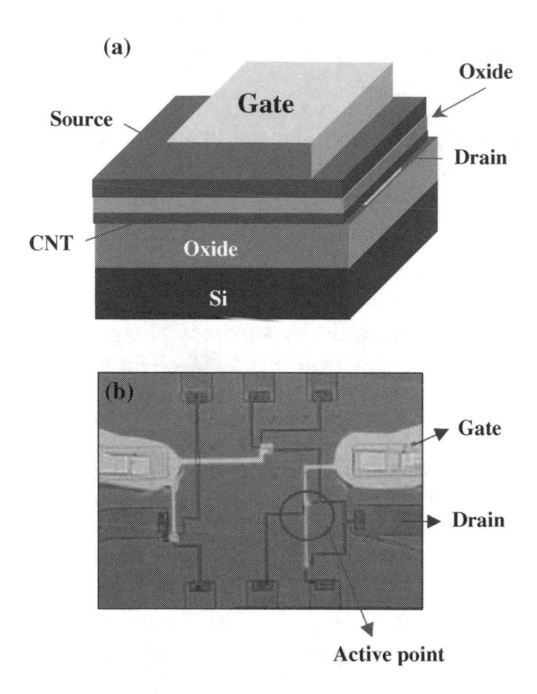

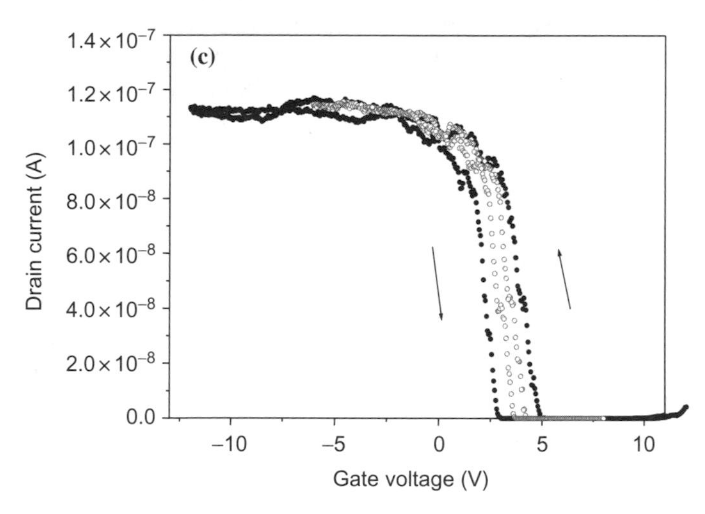

Figure 2.9 (a) Schematic diagram of a CNT-based non-volatile memory with ONO charge trap. (b) Image of a CNT-FET memory in which the electrode was patterned by electron emission lithography. (c) Drain current versus gate voltage of a CNT memory with top gate and an ONO layer. Drain current as a function of gate voltage and source–drain bias of −0.9 V. The maximum applied gate voltages in a sweep loop are 8 V and 12 V

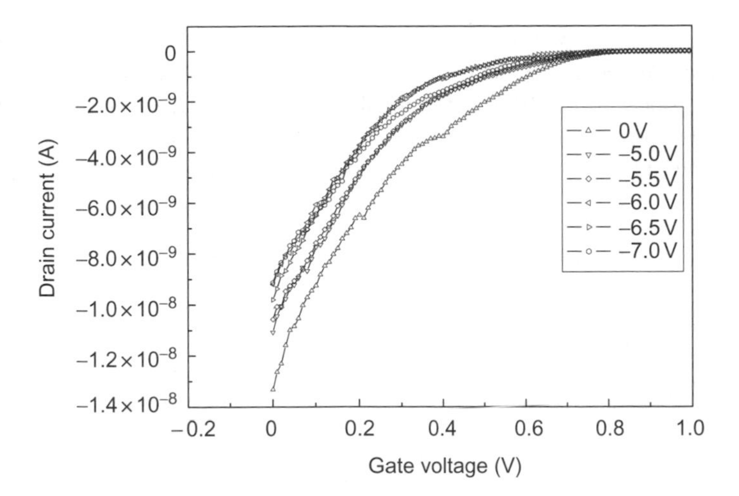

Figure 2.10 Drain current versus gate voltage of a CNT memory after charging the ONO storage node. A positive voltage pulse of duration 100 ms was applied to the gate, ranging from 0 to 7 V relative to the grounded source; the drain was maintained at −0.9 V

dielectric constant of 3. The calculated electric field for $V_g = 5$ V is 970 V/μm, which is high enough to produce Fowler–Nordheim tunnelling. Furthermore, supposing that the tunnelled charges flow along the electric field line, they will be trapped in the nitride layer, depending on the field intensity calculated by induced charge distribution. Our calculation shows that 70% of total tunnelled charges, which corresponds to the full width at half maximum (FWHM) of peak

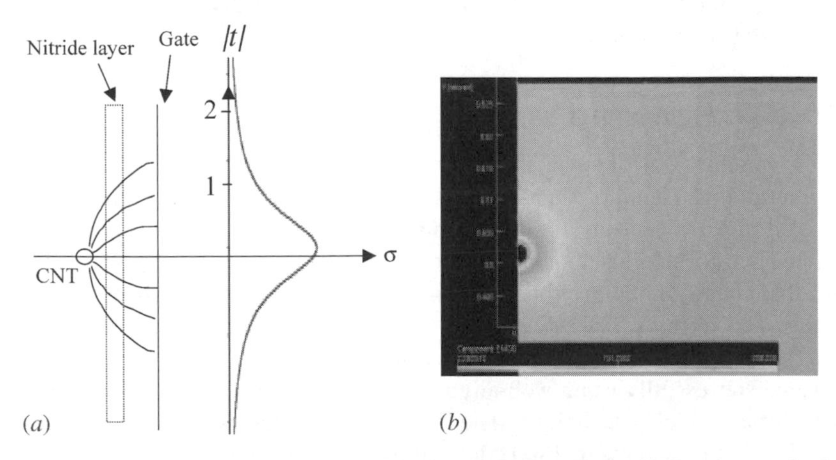

Figure 2.11 (a) Schematic diagram of the electric field between CNT and gate. Surface-induced charge density as a function of the arbitrary unit of CNT-to-gate electrode is shown on the right-hand side. (b) Simulation result of electric field distribution near a CNT. The region of maximum electric field is located around the CNT

surface charge density, will arrive within the 14 nm thick nitride layer on top of the tunnel oxide. The quantized state at room temperature is reported when the size of the quantum dot is below 10 nm. The localized charge distribution may be induced in the nitride layer due to the localized high field distribution of CNT. The trapped charge in the localized area may be able to diffuse to the uncharged area, but the

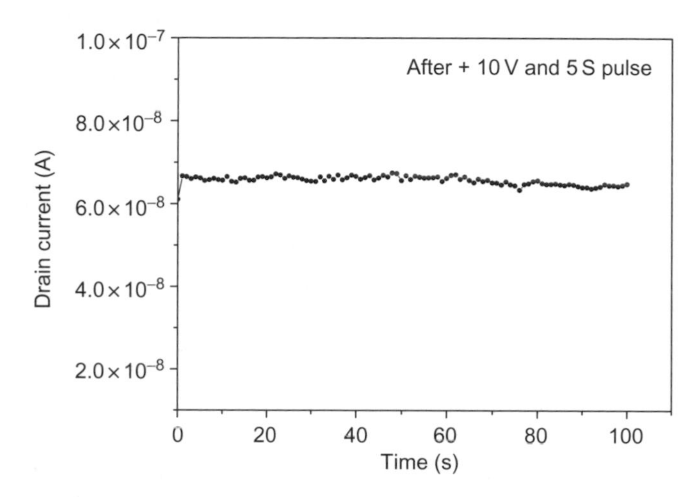

Figure 2.12 The measured drain current as a function of time for 100 s. The current was unchanged with time

current was unchanged with increasing time (Figure 2.12). This suggests that each trap site containing charge in the ONO layer of the CNT memory acts like a quantum dot for flash memory.

2.3.6 CNT Field Emission Display

Carbon nanotubes are known to be the best available field emitters. Their high aspect ratio, high chemical stability, high thermal conductivity and high mechanical strength are advantageous for field emitter applications. Carbon nanotube field emitters have considerable potential to be applied in emissive devices, including flat panel displays, cathode-ray tubes, backlights for liquid crystal displays, outdoor displays, and traffic signals. Following the first field emission from nanotubes in 1995, a prototype FED was demonstrated in 1999 [16–18]. CNT FEDs have been fabricated successfully using well-aligned nanotubes produced by paste deposition and a surface rubbing technique. The fabricated displays were fully scalable and showed a high brightness of 1800 cd/m^2 at 3.7 V/μm from the green phosphor. The fluctuation of the current was about 7% over a 4.5 in cathode area.

Figure 2.13(a) shows SEM images of SWNTs. Figure 2.13(b) shows TEM images of as-fabricated SWNTs. Bundles of SWNTs with diameters of about

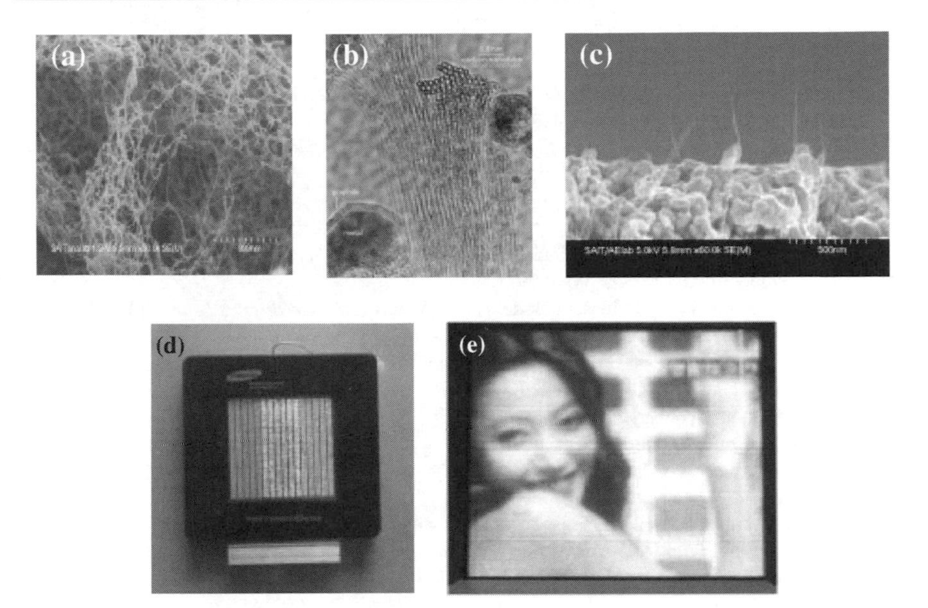

Figure 2.13 (a) SEM image and (b) TEM image of single-wall CNTs used for a FED cathode. (c) SEM image of CNTs attached onto the metal electrode after screen printing and surface treatment. (d) The first CNT FED image reported in 1999 and (e) a moving image on a CNT FED

1.4 nm are clearly seen. Metal particles were attached at the edge of the SWNT bundles. A very uniform and stable emission image over the entire display panel was obtained. Figure 2.13(c) shows a cross-sectional SEM image of a CNT cathode. It clearly shows that CNT bundles are firmly adhered onto the metal electrode and aligned mostly perpendicular to the substrate. The density of CNT bundles from the SEM measurements was 5–10 μm^{-2}, about 100 times larger than the typical density of microtips in conventional Spindt-type FEDs [19, 20]. SAIT and Samsung SDI have been involved in making large CNT-FEDs by using CNT paste printing technology. In the near future it might be possible to fabricate large panels of over 40 in with high uniformity.

Acknowledgement

The authors gratefully acknowledge the financial support provided by the National Program for Tera-level Nanodevices of the Ministry of Science and Technology as one of the 21st Century Frontier Programs. WB also thanks Byoung-Ho Cheong, Eunju Bae, Ju-Jin Kim and Young-hee Lee for their work on the carbon nanotube project.

References

1. Jhinhwan Lee, H. Kim, S.-J. Khang, G. Kim, Y.-W. Son, J. Ihm, H. Kato, Z. W. Wang, T. Okazaki, H. Shinohara and Young Kuk, *Nature* **415** (2002) 1005.
2. Taeghwan Hyeon, Su Seong Lee, Jongnam Park, Yunhee Chung and Hyon Bin Na, *Journal of the American Chemical Society* **123** (2001) 12798.
3. Haeseong Lee, Seung Ae Kim, Sang Jung Ahn and Haiwon Lee, *Applied Physics Letters* **81** (2002) 138.
4. R. Saito, M. Fujita, G. Dresselhaus and M. S. Dresselhaus, *Applied Physics Letters* **60** (1992) 2204.
5. J. W. G. Wildoer *et al.*, *Nature* **391** (1998) 59.
6. R. Martel, T. Schmidt, H. R. Shea, T. Hertel and P. Avouris, *Applied Physics Letters* **73** (1998) 2447.
7. P. G. Collins, M.S. Arnold and P. Avouris, *Science* **292** (2001) 706.
8. Won Bong Choi, Byung-Ho Cheong, Soodoo Chae, Eunju Bae, Jo Won Lee, Jae-Ryoung Kim and Ju-Jin Kim, *Applied Physics Letters* **82** (2002) 275.
9. Won Bong Choi, Byoung-Ho Cheong, Ju Jin Kim, Jaeuk Ju, Eunju Bae and Gwangsuk Chung, *Advanced Functional Materials* **13** (2003) 80.
10. Eun Ju Bae, Kwang Seok Jeong, Jae Uk Chu, In Kyeong Yoo, Won Bong Choi' Gyeong-Su Park and Seahn Song, *Advanced Materials* **14** (2002) 277.
11. Keun Soo Kim, Dong Jae Bae, Jae Ryong Kim, Kyung Ah Park, Seong Chu Lim, Ju-Jin Kim, Won Bong Choi, Chong Yun Park, Young Hee Lee, *Advanced Materials* **14** (2002) 1818.
12. A. Bachtold, P. Hadley, T. Nakanishi and C. Dekker, *Science* **294** (2001) 1317.
13. A. Javey, Q. Wang, A. Ural, Y. Li and H. Dai, *Nano Letters* **2** (2002) 929.
14. H. Bachhofer, H. Reisinger, E. Bertagnolli and H. von Philipsborn, *Journal of Applied Physics* **89** (2001) 2791.
15. V. A. Gritsenko, Hei Wong, J. B. Xu, R. M. Kwok, I. P. Petrenko, B. A. Zaitsev, Y. N. Morokov and Y. N. Novikov, *Journal of Applied Physics* **86** (1999) 3234.
16. W. B. Choi, D. S. Chung, J. H. Kang, H. Y. Kim, Y. W. Jin, I. T. Han, Y. H. Lee, J. E. Jung, N. S. Lee, G. S. Park and J. M. Kim, *Applied Physics Letters* **75** (1999) 3129.
17. W. B. Choi, Y. W. Jin, H. Y. Kim, S. J. Lee, M. J. Yun, J. H. Kang, Y. S. Choi, N. S. Park, N. S. Lee and J. M. Kim, *Applied Physics Letters* **78** (2001) 1547.
18. Won Bong Choi, Young Hee Lee, Nae Sung Lee, Jung Ho Kang, Sang Hyeun Park, Hoon Young Kim, Deuk Seok Chung, Seung Mi Lee, So Youn Chung and Jong Min Kim, *Japanese Journal of Applied Physics* **39** (2000) 2560.
19. B. R. Chalamala *et al.*, *IEEE Spectrum* **35** (1998) 42.
20. J. M. Kim, H. W. Lee, Y. S. Choi, N. S. Lee, J. E. Jung, J. W. Kim, W. B. Choi, Y. J. Park, J. H. Choi, Y. W. Jin, W. K. Yi, N. S. Park, G. S. Park and J. K. Chee, *Journal of Vacuum Science and Technology B* **18** (2000) 888.

3

Nanotechnology in Europe

Otilia Saxl
UK Institute of Nanotechnology

3.1 The Case for the European Research Area

At the Lisbon summit in March 2000 Mr Phillippe Busquin, the European Commissioner for Research, announced that his aim was to create a European research area that would become the most competitive knowledge-based economy in the world by 2010. Even as he made this statement, he was aware that this was a very ambitious target. Using current patent data in the advanced technology sector as an indicator, Europe falls behind both the US and Japan, holding only 9% of patents at the US office in comparison to 57% and 22% for the US and Japan, respectively. These ratios look a little better at the European Patent Office, where both Europe and the US hold 36% and Japan holds 22%.

If Europe is recognised for its high standards of research, so the real problem lies in industrial innovation or technology transfer. European performance in the field of innovation is still too limited, and there is much work to be done before Europe can compete, according to the vision of Mr Busquin. If future research can be moulded to suit the technological requirements for innovation, its impact will be stronger. This kind of research programme will have a greater impact if it is organised at the European level to suit the requirements of globalisation and the emergence of new markets. This is not only true for applied but also for fundamental research. A strategic pan-Europe an research programme could pave the way for the development of novel products and services that would lead to the realisation of Mr Busquin's target.

Nanotechnology: Global Strategies, Industry Trends and Applications Edited by J. Schulte
 ISBN: 0-470-85400-6 (HB)

The work programmes of the present EU framework programme are in support of this goal and place emphasis on socio-economic impact, sustainable development, and reduction in energy usage. Past programmes have helped to develop a culture of scientific and technological cooperation between different EU countries. This sixth framework programme (FP6) has been redefined and streamlined to achieve a lasting impact and a greater coherence at the European level and to focus efforts on fewer priorities. These priorities are in fields such as the life sciences and biomedical technologies, advanced IT, cognitive sciences, nanotechnology, and importantly, at their intersection. Europe recognises it will only obtain a share of the new, developing markets if it builds up its research sector in those key priorities by reinforcing a more intense collaboration between the academic sector and industry.

Nanotechnology is currently receiving €400 million of public funding each year in Europe. If regional and industry funding is added to this figure, the total could be as much as €1.2 billion annually. In general, the new framework programme will result in two to three times as much funding for nanotechnology compared with the previous investment by the EU.

3.2 Why Nanotechnology?

Why is it that nanotechnology has being selected for this strategic research push in Europe? In part, Europe is following the lead of the US and Japan, where government funding for nanotechnology research has increased year-on-year for over a decade. Internationally, high-profile funding for nanotechnology followed the announcement of dedicated funding for cross-disciplinary research under the National Nanotechnology Initiative by Bill Clinton in January 2000. The Japanese followed by establishing the Expert Group on Nanotechnology under the Japan Federation of Economic Organizations (Keidanren) Committee on Industrial Technology to examine nanotechnology. This expert group confirmed the importance of nanotechnology and encouraged the development of research programmes (www.jef.or.jp/en/jti/200109_010.html Japan Economic Foundation).

So why nanotechnology? Government and industry view nanotechnology as offering tremendous economic opportunities by optimising the life cycle of materials and products, increasing productivity and, critically in Europe, breaking the link between environmental impact and economic growth. As well as economic impact, nanotechnology promises many exciting opportunities to dramatically enhance healthcare and the quality of life,. The ability of scientists to visualise and control the behaviour of materials and at the nanoscale is providing them with the tools to develop novel products. At the nanoscale, materials contain novel and unexpected properties providing a real opportunity to create ‘smart’ materials that result in products with completely new functions. These products should be less resource- and energy-intensive to produce. Through technology at the nanoscale, the vision is that manufacturing will become steadily cleaner and greener, and products will be cheaper and have more functionality. There are few areas where nanotechnology is not forecast to make an impact. Here are some examples:

- *Healthcare*: nanotechnology will lead to the development of medicines tailored to the individual, and accurately delivered to the site of infection or disease; there will be new and better surgical techniques, often using robotics; retinal implants, cochlear implants and the ability to rejoin damaged nerves will be commonplace.
- *Energy*: energy will be used more efficiently through the development of smart buildings that respond to the environment; lighter and stronger materials for vehicles which will reduce production costs and decrease fuel consumption, cheap and versatile means for generating energy (polymer photovoltaic cells) and for storing it (by using carbon nanotubes for hydrogen storage).
- *Environment*: nanotechnology is leading the way in the development of molecular sieves for clean-up and to reduce pollution, and new disposable sensors to monitor pollutants in air and water.

These potential benefits are reflected in the commitment to nanoscale research by the world's major economic powers.

The field of nanotechnology is now growing dramatically and resources are still needed, especially to create the necessary infrastructure, including well-equipped facilities for research and prototyping, as well as a trained workforce. For Europe to really benefit from nanotechnology, several fundamental issues need to be addressed. Firstly, that scientists in academia and industry are in a position to increase their basic understanding of the nanoworld, by giving them access to funding and facilities so they can create new materials, devices and processes; so that they can work on integrating nanocomponents in micro and macro applications and establish new tools and techniques to enable industrial-level manufacture. There are also many further challenges, including the development of standards and metrological techniques for ensuring quality control and repeatability, and the need to be alert to and respond to any public awareness or safety issues. These challenges call for the creation of Europe-wide industry/research alliances of a significant scale for which a cohesive Europe-wide approach is demanded.

A major problem is the shortage of scientists in Europe, already discernible today. There are 5.1 researchers for every 1000 active persons in the European Union (EU), compared to 7.4 in the United States and 8.9 in Japan. This gap is even wider if the number of researchers in companies is considered: 2.5 per 1000 employees compared to 7 in the United States and 6.3 in Japan. This is especially true in the area of nanotechnology. In recent years there has been an increase in the number of postgraduate qualifications and summer schools offered in nanotechnology across Europe to address this issue. FP6 aims to address this problem by providing more money for education and exchange of scientists.

Many of the member states of the EU have dedicated nanotechnology research programmes (described below). In Europe, proximity is essential to the development and dissemination of innovations. It is hoped to encourage nanotechnology clusters based on geographical location to foster further development. The cluster Baden-Württenberg, Rhônes-Alpes, Lombardy, Catalonia brought together 'the four European driving forces' and gave rise to the common promotion of a 'Bio-Valley' in the biotechnology sector; this is viewed as a role model for nanotechnology development.

3.2.1 EU Funding History

Under the fourth framework programme (1994–1998), the EU spent approximately €30 million per annum on nanotechnology projects. Under the fifth framework programme (1998–2002), the EU spent €45 million per annum on nanotechnology projects. There were opportunities for nanotechnology funding in three thematic programmes:

- *Life sciences programme*: this looked at quality of life and management of living resources.
- *Information society technology programme*: nanoelectronics research projects were funded under the Future and Emerging Technologies (FET) Nanotechnology Information Devices (NID) activity.
- *GROWTH programme*: nanotechnology projects were funded under the generic activity of materials. Of the total budget for this activity of €65 million, €25 million were dedicated to nanotechnology projects. The EU programme on competitive and sustainable growth includes industry-oriented research in nanostructured materials.

To foster scientific and technological cooperation between the EU and the US government, the GROWTH programme entered into an agreement with the US National Science Foundation (NSF) in support of materials research. This enabled selected US researchers to join European consortia as participants in EU-funded research and technology development (RTD) activities under GROWTH's generic activity on materials, with the NSF providing support for US participants. A valuable outcome of this collaboration has been a series of joint EU/NSF workshops, organised on both sides of the Atlantic, to allow for fruitful exchange between scientists. This will remain an important area for collaboration during FP6. Negotiations are currently under way for the signing of further implementation agreements with Russia and China.

In FP5 more than 40 nanotechnology-based projects were funded. One such project was ROBOSEM, focusing on the development of a nanorobot system for a scanning electron microscope and nanotesting applications. This robot features sensor feedback from video cameras and force microsensors, and a virtual reality representation of the working environment.

The MONA_LISA project investigated new nanostructures for field-effect transistors (FETs), made using unconventional parallel lithography and growth techniques. This technique reduces defects, improves performance and brings energy savings. Both of these projects are still ongoing. FETs are vital electronic components.

The recently completed MicroChem project also made important progress in the production of laboratory-on-a-chip analysers for cost-effective monitoring of water purity on the basis of chemical analyses involving only nanolitre quantities of liquid. This allows much more thorough monitoring of the quality, and thereby the security, of effluents and critical potable water resources.

Projects based on nature's idea of building materials from the bottom up include NANOMAG, which explored new corrosion-resistant coatings for lightweight magnesium alloys. The objectives are to eliminate other processes involving carcinogenic compounds and to achieve a more widespread use of a hitherto unstable alloy that has inherent sustainability properties. These coatings are allowing magnesium alloys to be used increasingly in automotive and aerospace construction.

NEON used nanocrystals in the fabrication of new electronic memory devices. These crystals are produced by ion beam synthesis or deposition techniques. Use of nanocrystals helps to increase information storage densities while reducing power demand and even the amount of material.

The CARBEN project hinged on the development of an industrial-scale system for manufacturing carbon nanostructures that have a more controllable porosity, and can give a fully active surface area 10 to 100 times greater than for the graphite-like material currently used. Expected applications are in supercapacitors with high energy and power densities for use in trains and other electric vehicles as well as in a unique regenerative fuel cell (RFC) technology for bulk electric storage.

Nanocomposites with tailor-made electrical, magnetic or chemical properties are the basis of the NANOPTT project, made by filling nano-sized holes in polymer membranes with various combinations of metals or other polymers. Such a material can be used as screening to shield microwave ovens and mobile phones or as the active sensors in 'artificial noses' and miniaturised lab-on-a-chip devices for detecting biochemical reactions.

As a project cluster, NANOTRIB established synergies between six pre-existing projects, working in the field of nanoscale lubrication films and low-friction surfaces. This grouping involved a total of 60 partners from 16 countries, including 24 SMEs, backed by an investment of €16 million, of which the commission provided half. The projects are multidisciplinary and addressed multisectoral applications, from metal forming and machine tools to automotive engines, wind turbines and satellite mechanics. In addition, each of its constituents makes a contribution to sustainability by minimising the use of materials through an enhancement of performance at the nanoscale, by optimising the use of renewable organic-based lubricants, and by seeking to extend product lifetimes and reduce energy consumption.

The NANOTRIB cluster includes

- MICLUB: processing structured hard coatings for microlubrication.
- LUBRICOAT: examines environmentally friendly lubricants and low-friction coatings.

- HIDUR: development of nanocomposite coatings to improve competitiveness and conserve the environment.
- RIBO: nanostructured coatings for engineering tribological applications.
- NANOCOMP: nanocomposite wear-resistant and self-lubricating PVD coatings for tools and components. PVD is physical vapour deposition.
- SMART QUASICRYSTALS: tailored quasicrystalline surface layers for reduced friction and wear.

In the latter part of FP5, the GROWTH programme anticipated the increasing interest in nanotechnology and the need to create a broad thematic network. It funded a pan-European thematic network launched in July 2002, called NANO-FORUM (www.nano.org.uk). Nanoforum will continue throughout the four years of the FP6 programme. Its broad frame of reference provides a basis for raising awareness, supporting and encouraging the adoption of new nanotechnologies, and facilitating the development of new industrially oriented nanotechnology research across Europe. Another major activity will be the wide dissemination of information, to the public as well as science and industry, using the media, its website and special interest groups.

3.3 Nanotechnology in FP6

The sixth framework programme (FP6) will span the period 2002–6, replacing the current programme FP5.

In FP6 nanotechnology has been highlighted as a key area for European development and, unlike in FP5, has become a priority area. The wide potential spread of applications for nanotechnology means that its impact will be felt across virtually the whole programme. Priority 3, on nanotechnologies and nanosciences, knowledge-based multifunctional materials, new production processes and devices (NMP), is the main vehicle for research in this area.

By bringing together nanotechnologies, materials science and manufacturing as well as other technologies based for example on biosciences or environmental sciences, work in this area is expected to lead to real breakthroughs and radical innovations in production techniques and consumption patterns. The intention is to promote the transformation of today's traditional industries into a new breed of interdependent high-tech sectors. The objective is to faciltate real industrial breakthroughs and promote sustainable development across activities ranging from basic research to product development in all technical areas from materials to biotechnology. Funding will be around €1.3 billion.

Prior to the construction of the work programme, a call for proposals was issued to determine the readiness, understanding and interest of the RTD community in submitting proposals and to provide the researchers with the opportunity to have some input into the work programme itself.

This resulted in over 1670 expressions of interest (EoIs): 396, or 24%, were considered 'mature and promising' and 882, or 52%, were considered to be relevant but did not demonstrate sufficient breakthrough research; the remainder were not considered appropriate. The EoI exercise acknowledged the strength of nanosciences in Europe and confirmed the importance in translating this research into a real competitive advantage for European industry.

The key areas identified were

- mastering processes and developing research tools, including self-assembly and biomolecular mechanisms and engines;
- activity at the interface between biological and non-biological systems as well as surface-to-interface engineering for smart coatings;
- nanoscale engineering to create materials and components, including the development of instrumentation for use at the level of atoms and molecules;
- chemistry, catalysis and reactivity and new eco materials;
- engineering support for materials development leading to new materials by design, e.g. biomimetic and self-repairing materials in the context of sustainability;
- new processes and flexible intelligent manufacturing systems based on nanotechnology and new materials;
- systems research and hazard controls to allow radical changes in the basic materials industry;
- optimising the life cycle of industrial systems products and services;
- Integration of nanotechnologies for improved security and quality of life, especially in the areas of healthcare and environmental monitoring.

3.3.1 Nanomaterials Challenges

The challenge in the field of materials research is to create smart materials that integrate intelligence, functionalities and autonomy. Such materials will not only provide innovative answers to existing needs, but also accelerate the transition from traditional industry to high-tech products and processes. Knowledge-based multifunctional materials are seen as an area that will contribute towards high value-added industries and sustainable development. It was felt that the strong research in this area could be translated into a competitive advantage for European industries. A further aim of FP6 funding is to promote the uptake of nanotechnology into existing industries such as health and medical systems, chemistry, energy, optics, food and the environment. Here are some other areas where nanotechnology will make an impact.

3.3.1.1 Food Quality and Safety

- The development of reliable traceability methods and systems to establish the origin of foodstuffs or their modes of production, all the way from farm to fork.
- New and more sensitive sensors for detection of health and environmental risks.

3.3.1.2 Genomics and Biotechnology for Health

- Technology development for exploitation of genetic information specifically in the area of high precision and sensitivity for functional cell arrays.
- Improved drug delivery systems.

3.3.1.3 Information Society Technologies

- Manufacturing, products and services engineering in 2010.
- Micro- and nanosystems.

3.3.1.4 Emerging Priorities

- Crime prevention and security for the people of Europe.
- Sustainable energy production.

3.3.2 COST: Cooperation in the field of Scientific and Technical Research

Founded in 1971, COST is an intergovernmental framework for European cooperation in the field of scientific and technical research, aimed at the coordination of nationally funded research at the European level. COST actions include basic and pre-competitive research as well as activities of public usefulness. COST has 33 member states, 1 cooperating state and 9 states with participating institutions. The latter are non-European countries. COST actions run for about four years. COST coordinates nationally funded research worth €1.5 billion per year. €60 000 is available per action for coordination.

There are 15 COST actions dealing with nanosciences and technologies. Two COST actions in telecommunications applications deal with nanotechnologies, COST 265 and 268. In the first, researchers are developing new measurement techniques for active and passive fibre standardisation. In the second, the work is focusing on wavelength-scale photonic components. Seven COST chemistry actions involve nanotechnology. They are supramolecular chemistry (D11); computational chemistry of complex systems (D9); functional molecular materials (D14); interfacial chemistry and catalysis (D15); polymers, etc., via metal catalysis (D17); metalloenzymes and chemical biomimetics (D21); and protein–lipid interactions (D22).

Four COST materials actions deal with nanotechnology. COST 523 is in nanostructured materials; COST 527 deals with plasma polymers. In COST 528 they deposit chemical solutions on thin films. COST 525 is in electroceramics, where they engineer grain boundaries. Two COST physics actions involve nanotechnology. P1 is in soft condensed matter, and P2 is involved in mesoscopic electronics. More information on these COST projects can be found at http://cost.cordis.lu/src/home.cfm.

3.3.3 Supporting SMEs

The European Union is currently supporting co-operative research projects (CRAFT actions) aimed at enabling groups of small and medium-sized enterprises (SMEs) from different countries to submit proposals to research centres or universities on research they might require for technological purposes. This type of encouragement has given rise to some remarkable success stories, such as the implementation of clinical tests for an artificial cornea, developed by the company Cornéal based near Annecy. There is still a need for more involvement from SMEs. To encourage this, 15% of funding will be allocated to SME involvement in FP6. This breaks down to €1.7 billion, excluding SME-specific projects such as CRAFT and collective research projects. Consultation between scientists from academia and industry and the EU has been the basis of strategies for the growth of opportunities in funding nanoscience and nanotechnology.

3.3.4 Previous National Policies on Nanotechnology

During the 1990s, European countries, including Germany, France, the Netherlands, Spain and the UK, organised and conducted national forecasting activities to identify priorities for technology policies. The aim was to systematically accumulate knowledge on those technologies likely to be extremely influential in the future. Here are some examples:

- *Germany*: A Delphi study, resulting in the report 'Technologies of the 21st Century' (Ministry for Education and Research, 1993), culminated in a 1988 report by VDI-TZ (Technology Centre of the German Engineers Society) entitled 'Opportunities in the Nanoworld'.
- *UK*: A National Initiative on Nanotechnology was established in 1986, followed by a LINK programme of collaborative research in 1998. A report on nanotechnology by the Parliamentary Office of Science and Technology (POST) was published in 1996 and the Institute of Nanotechnology was founded in 1997.
- *Netherlands*: In the 1990s it commissioned studies on nanotechnology from the Dutch Foresight Committee (OCV, 1995) and the Study Center for Technology Trends (1999).
- *Europe*: Scientific and Technical Options Assessment (STOA, 1996) published a study on nanotechnology for the European Parliament and the Institute for Prospective Technological Studies (IPTS) also published papers on nanotechnology for the European Commission in 1997.
- *World*: The World Technology Evaluation Center (WTEC) has published an annual reports on nanotechnology since the first one in 1999; the National Nanotechnology Initiative (NNI) was announced in 2000.

Today most industrialised European countries now have government-supported major nanotechnology research and development initiatives, but it comes as no surprise that the countries with the largest economies are the most active. Up till now, Germany has been the most active EU country in nanotechnology, where the federal government has funded competence centres in nanotechnology-related areas, following the successful example of competence centres in biotechnology. The Nanonet competence networks bring together public research institutes, industries and SMEs to collaborate on relatively application-oriented research topics, stimulating technology transfer.

The Swiss national programme TOPNano21 is similar to NNI in the US. The French government has also set up a structure where it attempts to centralise funding for micro- and nanotechnology research. The UK has created interdisciplinary research centres at Oxford and Cambridge, and has invested in other universities. It is predicted that UK funding for nanotechnology will overtake that of other European countries in gross terms.

Austria, Ireland and the Netherlands are putting programmes in place, and the Swiss per capita spending on nanotechnology is still the highest in Europe by a considerable margin. Many of the smaller European countries, in particular the southern countries, are not concentrating on nanotechnology but are focusing on the more traditional industries, except for Italy, which received more funding for nanoprojects than any other country in the last round of FP5.

Often the boundary between nano- and microsystems technologies is weak. Programmes are frequently termed 'micro- and nanotechnology' (e.g. in France and the Netherlands). Nanobiotechnology is also becoming popular, but here too the boundary with mainstream biotechnology is fuzzy (e.g. biochips are claimed by both). With no consensus on a definition of nanotechnology, different programmes are difficult to compare. Other countries fund nanotechnology research through their generic research budgets or through other dedicated programmes (e.g. in materials or microelectronics).

Economically, a sensible strategy for nanotechnology is to focus on niche markets, where there exist no commercially available, yet cheap, established technological solutions. Which niche markets are relevant for nanotechnology? In Europe our strengths lie in the healthcare and life science markets. An example is lab-on-a-chip technology for cheap and easy-to-use diagnostics. The Institute of Nanotechnology in the UK is a promoter of this strategy. VDI-TZ German and government studies which prepared the ground for the federal government's competence centres on nanotechnology, looked in detail into the potential of nanotechnology for different key industrial sectors, such as

- medicine, pharmacy, biology;
- precision engineering, optics, analytics;
- chemistry and new materials;
- electronics, information technology;
- automotive.

Technological and economic developments are moving rapidly today, and many competitors are working towards the same applications for niche markets, as well as for more mature competitive markets. For the EU and national policy makers, the societal relevance of research is not restricted to economic gains such as employment and competitiveness of EU industries. These decision makers fund research with taxpayers' money, which means that their priorities include better healthcare, sustainable development, better use of renewables, and so on. At this stage, one can foresee that nanotechnology is likely to contribute to improved medical therapies, and there are arguments that nanotechnologies will contribute to sustainable development through miniaturisation, leading to reductions in the quantity of materials needed, the creation of novel materials for transport, reducing fuel requirements, and the production of cheaper renewables such as more efficient solar cells. It is difficult to quantify the effect accurately.

The EU Innovation programme funded a European consortium on micro- and nanotechnologies and markets. Its aim is to assist SMEs in participating in European research projects under FP6. It provides support for scientists wishing to use large-scale research facilities in other EU member states. Currently, the programme covers access to five large-scale research installations for nanoscience, in Israel, Germany, France, Spain and Ireland. The Weizmann Institute in Israel has opened the doors of its Center for Submicron Research on Semiconductors to European researchers. Researchers may also use the equipment of the Institute of Microtechnology (IMM) in Mainz, Germany. IMM specialises in microfabrication methods for applications in fields such as sensors, actuators, biomedical applications, microreactors and micro-optics. Researchers can apply to use the clean room at the microelectronics facility and related services (MICROSERV) at CNM in Barcelona. MICROSERV is upgrading its equipment with the acquisition of new instruments such as an e-beam lithography system. The National Nanofabrication Facility at NMRC in Cork, Ireland, is also accessible to other Europeans.

3.4 National Funding Policies

3.4.1 Austria

In Austria the work programme of the federal government provides for a number of far-reaching reforms for the areas of science, research and technology. One of the priority objectives is to increase the research quota to 2.5% of GDP by 2005. An intermediate target is to increase the quota to 2.0% of GDP by 2002. Although it does not appear to have a specific national nanotechnology programme, the Austrian Council has launched a number of initiatives to develop, strengthen and promote emerging technology fields for the future, and nanotechnology has been included.

The FWF (Fonds zur Föderung der wissenschaftlichen Forschung) does not make calls for specific nanotechnology but currently funds a limited number of projects. Here are some examples:

- New methods for quantum optics with clusters and molecules, at the University of Vienna. This project has possible applications in the measurement of

molecular properties or external fields. Such techniques may be useful in the preparation of extremely small and regular molecular structures which, in the long run, could be useful for surface chemistry or molecular data processing.
- UHV-AFM studies of nanodefect formation by slow multicharged ions at insulator surfaces, at the University of Vienna. The main goal of this project is to achieve sufficient clarity on whether coverage of insulator surfaces by small metal clusters would be a possible way for producing regularly structured PSI-induced surface nanodefects on insulator surfaces relevant for microelectronics and nanotechnological applications.
- Low dielectric constant polymers as novel charge electrets, at the University of Linz. This project aims to identify new charge electrets. The results may lead to new applications in advanced electret devices. From a fundamental point of view, new insights are expected in charge trapping in polymers, as well as in the characterisation of softening and glass transitions in non-polar thermoplastic and thermoset polymers. The results may also be applicable to materials in other fields, such as glues and dental resins.
- Membrane trafficking of activated intermediates in the glycosylation pathway of S-layer proteins, at the Zentrum für Ultrastrukturforschung, Universität für Bodenkultur Wien. Two projects in this area of nanobiotechnology are being funded in the general context of planned future applications of recombinant bacterial S-layer neoglycoproteins in biotechnology, medicine and nanotechnology.

A mapping of RTD competencies and potential in Austria is soon to be undertaken. The Austrian Council will control and monitor programmes and initiatives recommended for financing by the special R&D funds.

Austria's research expenditure is currently average among members of the Organisation for Economic Cooperation and Development (OECD), with a current rate of about 1.9% (measured against the gross domestic product). In 2000 its R&D expenditure was estimated to be slightly above €3.634 billion (Sch 50 billion). 32% of R&D expenditure was accounted for by the federal state, which is significantly higher than the international average share of public R&D funding. The enterprise sector, on the other hand, was slightly below the EU average, accounting for approximately 40%. European integration has contributed significantly to international groups using Austria's advantages as a location. Currently, about 21% of research expenditure in Austria is financed from abroad, in particular by European enterprises which have chosen Austria as their research location.

3.4.2 Belgium

Belgium is unique in the way it funds science with both research agencies and universities divided into Flemish- and French-speaking organisations that are relatively autonomous. The IWT, the Flemish Institute for the Promotion of Scientific-Technological Research in Industry, coordinates nanotechnology projects. In the

STWW (strategic technologies for welfare and well-being) programme, IWT also funds projects on nanopowders and nanostructured materials. Half of IWT's budget of €3 billion a year is devoted to R&D in information and communication technology (ICT), including telecommunications, software and microelectronics. Different formulae exist. One encourages the introduction of R&D projects irrespective of their subject, with selection based on general R&D quality. The EUREKA projects, Tessi and Media belong to this category. Another is more top-down, with the government fixing a number of strategic themes for industrial R&D projects. The main topics in this approach include multimedia and communication technology (ITA programme II, with a budget of roughly €1 billion), and microelectronics and embedded software (both part of the EUREKA programme).

In 1982 the Flemish government set up a comprehensive microelectronics programme to strengthen the microelectronics industry in Flanders. This programme included the establishment of IMEC, a laboratory for advanced research in microelectronics headquartered in Leuven, which has now become Europe's largest independent research centre in this field. IMEC concentrates on design technology for integrated information and communication systems; silicon process steps and modules, silicon processes; nanotechnology, microsystems, components and packaging; solar cells; and advanced training in microelectronics. Its revenue of more than €130 million is derived from agreements and contracts with the Flemish government and companies, equipment and material suppliers and semiconductor and system-oriented companies worldwide, the EU, MEDEA+ and ESA. Work in nanotechnology is funded by Belgian science foundations, the EU, FET and MEL-ARI programmes, and bilateral industrial contracts. At the Katholieke Universiteit Leuven (KU Leuven) several groups in the Laboratory of Solid State and Magnetism are engaged in nanotechnology research.

At the Technological Research Centre, VITO, they carry out contract research in materials. EMAT is the Centre for Electron Microscopy and Materials Science. It is based in the department of physics at the University of Antwerp (RUCA). They study nanotubes and particles using their electron microscopy. Collaborations include Agfa, De Beers, ICI, Phillips, Shell and Unilever. At Ku Leuven, CERMIN is a research centre in micro- and nanoscale electronics devices and materials. It was founded in 1998 to address the need for multidisciplinary research. The design of application-specific integrated circuits (ASICs) is a major research activity. The research also covers more fundamental areas, such as the modelling of fabrication processes, the realisation of novel devices, analogue integrated circuits, digital systems, methodologies for computer-aided design, and information security.

3.4.3 Denmark

Denmark aspires to be among the best in the world at developing and using new knowledge and technology, and in cooperation between companies and knowledge institutions by 2010. Denmark currently comes around the middle of nine prosperous OECD countries. (OECD R&D and MSTI database, May 2001). In 2002 the ministry published 'A Strategy for Technological Service 2002–2005'. This report is to be followed by an action plan for strengthening cooperation between companies and knowledge institutions, and a plan to improve the framework conditions for entrepreneurs and start-ups. Francois Grey (ex-MIC) estimated that by 2010 the Danish micro-industry will have a turnover of about DKr 10 billion (€1.3 billion) in micro- and nanotechnologies.

A good example of an innovative Danish centre is the cooperation between the National Micro- and Nanotechnology Research Center (MIC) at the Technical University of Denmark, the Authorised Technological Service Institute (DELTA) and the companies Danfoss, Grundfos, Microtronic and Capres. The cooperation is focused on the explosively growing microsystems technology which has already led to outstanding commercial results and perspectives:

- *Grundfos*: a pressure sensor chip for controlling pumps that is resistant to aggressive liquids and also cheap to make.
- *Mikrotronik*: a microphone the size of a pinhead. It contains electronics for handling signals and an amplifier. The microphone is primarily intended for use in hearing aids but is so cheap to make that it will also be of interest for other applications.
- *Capres*: an instrument that measures electrical surface resistance at a micro level on surfaces. It offers new possibilities for controlling thin films used in more and more contexts for electronic circuits.

MIC is an autonomous research centre on the campus of the Technical University of Denmark, with a staff of about hundred. Originally, MIC focused on sensors and hearing-aid development, a traditionally strong industry in Denmark. More recently there has been a move towards biotechnology and biochips, in conjunction with companies in Medicon Valley around Copenhagen. The strategy of the nanotechnology group at MIC is to develop new microtools for nanotechnology, using MIC's state-of-the-art clean room facilities for silicon microfabrication. Current projects in the group include micro-four-point probes, nano mass sensors, nano tweezers, and cantilever-based biosensors. MIC receives 60% of its funding from the Ministry of Research, the rest from industry projects and competitive grants. It also collaborates with multinational companies.

There are four other nanotechnology-related research centres in Denmark The first is COM (Communications, Optics and Materials). This is a telecommunication research centre where they work on optical Bragg gratings and photonic crystals. The second is ICAT, a research centre focusing on nanoscale functioning of catalysis, together with catalysis companies. The Danish Polymer Centre deals with micro and nanostructures of polymers. In the Department of Chemical Engineering, research activities of the aerosol laboratory include the manufacture of nanosized particles and nanoporous membranes by aerosol routes. Finally, at the Center for Atomic-Scale Materials Physics (CAMP) they study metallic nanostructures and their properties by a closely coupled experimental and theoretical approach.

As part of a political agreement in 2002, funds of DKr 25 million were allocated for 2003, with a further DKr 85 million being allocated for the years 2004 and 2005 for a larger national research effort within the areas of nanotechnology and nanoscience. From the total of DKr 25 million per year DKr 20 million are allocated for a national research effort in nanotechnology and nanoscience, and the remaining DKr 5 million are allocated for structured educational programmes in these areas. These funds are not allocated to specific topic areas but are to be used to strengthen Danish research within nanotechnology and nanoscience and to contribute to education within these areas. The Danish government is also looking for an economic return on its money, hoping it will contribute to international breakthroughs and new high-tech spin-off companies. Emphasis is given to enhanced collaboration across the Danish nanosector in order to prepare for participation in FP6.

3.4.4 *Finland*

Finland was one of the first countries to run a nanotechnology programme which spanned the years 1997–99. TEKES (Technology Development Centre Finland) supplied £4.3 million for 16 projects in the areas of

- nanobiology,
- self-organising structures,
- functional nanoparticles,
- nanoelectronics,
- biomaterials for information technology.

The Academy of Finland then started a related research programme in electronic materials and microsystems, running from 1999 to 2002, with a budget of Fmk 30 million (about $6 million). The themes were materials for silicon technology, new materials and interface effects, and development of devices and microsystems. At the moment, there is no national research programme on nanotechnology in Finland, although the National Microsystems programme does encompass some nanotechnology research.

The Academy of Finland is investing €5.1 million in 11 projects whose themes are

- materials for silicon technology,
- new materials and interface effects,
- development of devices and microsystems.

There are also several groups actively researching nanophysics, nanochemistry and nanomaterials science with emphasis on developing future products with added value using precision technologies. One such programme, called PRESTO, focuses on (microelectromechanical systems (MEMs), micro and precision fabrication, and micro and precision assembly.

The programme ELMO looks at miniaturising electronics and will run from 2002 to 2005 in a bid to maintain the Finnish market position in this area. It has an approximate budget of €100 million. The division is roughly €50 million from TEKES and €50 million from other sources, mainly industry.

VTT Electronics is also engaged in nanotechnology research. The main focus is on self-organised growth of semiconductor structures, quantum transport in mesoscopic silicon and silicon-on-insulator (SOI) structures and fabrication of nanoscale features using imprint techniques.

Other nanotechnology research takes place at the Helsinki University of Technology and includes work in optics at the Materials Physics Laboratory and research on nanostructured materials in the Laboratory of Polymer Physics. The physics department at the University of Jyväskylä has a nanotechnology research group working on nanophysics and nanofabrication.

3.4.5 France

In France, miniaturisation of microsystems technologies as well as nanoelectronics are the main foci of nanotechnology research. In 1999 CNRS created a centre of excellence in nanoelectronics 'from silicon to the molecule'. This centre is dedicated to developing the necessary components and materials for future information technology. It comprises 18 CNRS research labs; 44 other French labs also participate, including 3 industrial labs (CNET-CNS, STMicroelectronics and STM-FT) and 4 labs of the Atomic Energy Centre (CEA). The participating labs are listed at www.laas.fr/~temple/GDRnano/laboratoires.html.

Also in 1999 the French Ministry of Education and Research funded a web service in micro- and nanotechnology, to foster public–private research collaborations. This web service is called RMNT (Réseau de Recherche en Micro et Nano Technologies). The RMNT service acts as an intermediary between project proposers and the different French government institutions responsible for funding R&D. It selects innovative projects which address technological deficits and economic needs. It is only open to French proposers, and has an annual budget of FFr 70 million (about €10 million).

In 2000 the Ministry for Research installed the Action Concertée Incitative (ACI) in Nanostructures, to foster the development of new nanostructured materials for future semiconductor technologies. It is interdisciplinary, involving physics, engineering science, chemistry, biology and earth sciences (high pressure, extreme conditions). The four priorities are

- chemical and biological methods for self-organising materials fabrication;
- inorganic and molecular components, nanocontacts and interconnections;
- collective effects in nanostructures: magnetic, superconducting, ferroelectic;
- nano-optoelectronics, quantum dots and wires.

In 2000, FFr 10 million (about €1.5 million) were spent on 15 projects, each running two to three years. The funds come from the National Science Foundation. There were many more applications than could be funded.

In 2003 CNRS launched a programme for nanotechnology and materials development to fund research in about 40 physics laboratories and 20 chemistry laboratories. The government hopes that its actions will enable France to emulate Germany, which has doubled public investment over the past five years and taken over from Britain as Europe's leader in private investment. Priority has been given to information and communications sciences and technologies (ICST) where nanotechnology will make a big impact and support for research in ICST (photonics, nanotechnologies, cryptology, optoelectronics) is being actively promoted.

Much of the nanotechnology research in France is centralised at Minatec. Minatec was launched at the instigation of the CEA and the Institut National Polytechnique de Grenoble (INPGrenoble), bringing together 3500 people at Grenoble's scientific research centre. The site was already home to LETI, Europe's top centre for applied research in microelectronics and microtechnology, and its considerable resources. A total of €150 million will be invested in Minatec between 2002 and 2005 to fund the new infrastructure, in addition to the €250 million invested by CEA-LETI and INP Grenoble. Over the past 10 years, the microelectronics industry has invested €4 billion in the Grenoble-Isère area. Funding comes mainly from the Ministry of Industry and Research and the national research council, CNRS.

The main areas of nanotechnology focus at Minatec are

- physics (materials, magnetism, optics, quantum systems);
- instruments for observation at nanometric scale;
- chemistry;
- molecular electronics.

Nanostructured materials research projects are coordinated by the Réseau National Matériaux et Procédés (National Resource for Materials and Processes, RNMP). This network aims to foster innovative research, and to foster research-industry collaborations in five areas:

- conception, elaboration and characterisation of materials;
- process optimisation;
- surface treatment and assembly;
- controls of behaviour, durability and reliability;
- environmentally friendly processes and materials, recyclability.

RNMP evaluates proposals and recommends them for funding by the Ministry of Economy, Finance and Industry; the Ministry of Research; or ANVAR, which supports innovative SMEs (www.anvar.fr).

3.4.6 Germany

Currently Germany is very active in nanotechnology, with government funding of around €144 million annually plus a further €44 million coming from industry. It has a well-established infrastructure in the field, partly due to augmentation of its existing microtechnology infrastructure. The German Federal Ministry of Education and Research (BMBF) has recognised the significance of nanotechnology as a future technology and given large grants for nanoprojects in physics, chemistry, microelectronics and information-technology, as well as for biology and genetic engineering. Funding is provided for projects which lead to broad-based technological and economic exploitation, as well as for high-risk research. Germany sees energy engineering (gas cells, batteries, solar cells, etc.), environmental technology (materials cycles, clean-up) and information technology (memory, storage) as well as health and medicine as of vital importance. In 1998 BMBF established six competence centres to support research and advance the industrial application of nanotechnology. The role of these centres is to communicate with the public, link industry and university, and carry out training and further research. These six centres operate in the five key areas listed above.

- *Production and use of lateral nanostructures*: structuring technologies on a sub-50 nm scale for high-density data storage, electronic components with a structural width of under 50 nm, a new generation of electronic components, applications in biotechnology, etc.
- *Application of nanostructures in optoelectronics*: quantum effects in optoelectronic components for communication technology, environmental monitoring, CD players, laser TV, displays, etc.
- *Ultra-thin functional layers*: ultra-thin layers in electronic components, for sensors, implants and artificial skin, in X-ray optics or in abrasion and corrosion protection layers.
- *Functionality by means of chemistry, production and application of nanomaterials and molecular architectures*: new nanostructure materials for abrasion protection, for improving the characteristics of latex paint, in optics, for optimising catalysts or in glue processes, in pharmaceuticals or in auxiliary materials for electronics.
- *Ultra-precise surface processing*: smoothed surfaces with particularly precise shaping and little roughness, e.g. for high-precision optics for semiconductor technology.
- *Measuring and analysis of structures down to nanoscale dimensions*: measuring and analysis of surfaces and components for quality control and for reproducible mass production. Most prominent example of nano-analytics: scanning probe technologies.

Germany established these national competence networks to enable domestic manufacturers to commercialise nanotechnology. Large companies collaborate actively in the competence networks, and are very aware of what is being developed. The aim is to create jobs in innovative sectors in Germany and to protect the existing ones in a globally competitive market. The nanotechnology competence networks are coordinated in cyberspace on the German Engineer's Society's website at www.nanonet.de/. Large companies collaborate actively in the competence networks. An example is the Degussa project House Nanomaterials. This project started at the beginning of 2000 and is co-funded by the German Research Council (DFG) and Degussa. Degussa's staff members collaborate with researchers from nine German universities in developing new nanostructured materials. The DFG contribution covers research by scientists in the DFG priority programme 'Handling of highly dispersed powders'. The aim of the collaboration is to scale up the production of nanopowders from laboratory scale.

BMBF, in a joint initiative with the Physical and Chemical Technology Programme, also launched a funding programme in 2000. Entitled Nanobiotechnology, it will run for six years with a fund of €20 million. The funding stream is dedicated to multidisciplinary research projects related to

- analytical and characterisation processes,
- manipulation techniques for biological objects,
- reaction techniques for the analysis of structure–activity relationships,
- biological self-assembly for functional layers,
- cellular and molecular tools and machines.

The major goals of this activity are the rapid transfer of biological know-how into nanotechnology, the use of biological nano-objects in technical systems and the exploitation of nanotechnology in biotechnology and medicine.

BMBF has earmarked a total of €280 million for industrial collaborative, precompetitive and strategic academic research projects in the period 2002 to 2006 for research in optical science and engineering. At present, an estimated 15% of jobs in the German manufacturing industries depend on optical technologies. The research topics covered by the programme will be handled flexibly to reflect latest developments in the area of optical science and engineering. Here are some current research priorities:

- Innovative photon sources such as vacuum ultraviolet (VUV), extreme ultraviolet (EUV) and X-ray sources, ultra-short pulsed lasers, high-performance diode lasers, semiconductor lasers based on nanoscale materials, organic light-emitting diodes (LEDs), plasma technologies and quantum optics.
- Nano-optics, including ultra-precision optical engineering, EUV, X-ray and near-field technologies.
- Integration of optical technologies, including simulation, materials science, design, production, coating, assembly, metrology and functional testing.
- Biophotonics, i.e. applications of optical science and engineering to increase the knowledge of biological systems (e.g. innovative microscopy, spectroscopy, screening methods for cellular systems; advanced probes and markers).

- Metrology and testing, including metrology for micro- and nanoscale materials; process and quality control for EUV lithography; applications for the food industry and agriculture; innovative measuring and sensor technologies.
- Nanotechnology, including new (non-semiconductor) applications of VUV, EUV and X-ray sources; photonic materials processing for nanoscale operations.

Visit www.optischetechnologien.de (in German only) only and www.bmbf.de.

3.4.7 Ireland

Ireland is currently enjoying the benefits of a period of sustained economic growth and is currently facing new challenges. The industrial and economic future rests in Ireland becoming an innovation-driven economy. To address these challenges, the Irish government allocated almost €2.54 billion (IR£2 billion) in funding for research, technological development and innovation (RTDI) in the National Development Plan (www.ndp.ie) covering the years 2000 to 2006. This very substantial level of investment reflects the government's acceptance of the strong link between investment in the research and innovation base of the economy and sustained economic growth. These areas associated with nanotechnology were identified as funding priorities:

- advanced formulation or delivery and packaging systems, including smart drug delivery technologies;
- user interfaces, e.g. multisensory, wearable; virtual reality; artificial intelligence; human language understanding and synthesis;
- advanced materials, including biomaterials, smart materials and reusable or renewable materials;
- processing or fabrication of new and advanced materials, including exotic metals to prevent corrosion in the chemical industry, new polymers to prevent contamination in the food and healthcare sector, repair of turbines for aircraft;
- integration and miniaturisation technologies;
- exploitation of ICT and logistics;
- intelligent consumer energy products;
- prefabrication technology, use of robotics, mechanisation, tool technology.

Science Foundation Ireland (SFI) was established in 2000 to administer Ireland's Technology Foresight Fund. SFI is investing €635 million between 2000 and 2006 in academic researchers and research teams who are most likely to generate new knowledge, leading-edge technologies, and competitive enterprises in the fields underpinning two broad areas: biotechnology and ICT. There are many nanotechnology projects funded under this scheme. For example, at Trinity College Dublin (TCD), physicists are using the latest techniques to examine individual atoms in a biological sample in an attempt to grow miniature electronics atom by atom in a test tube. One project involves studying abnormal proteins called amyloid fibrils, formed in a biological process that might be useful for producing ultra-fine wires.

TCD is also developing a detailed atomic-level understanding of the methods used in silicon processing. This should dramatically improve the ability to control growth and etching of ultra-thin material layers The plan is to develop new protocols for assembling, fabricating and testing nanometre-scale device structures, to identify the essential building blocks for nanometre-scale devices, and to establish predictive rules for the assembly and performance of these devices.

At University College Galway, a set of spectacles is being developed that can give ‘super vision’. The approach, termed adaptive optics, uses novel electronics, computer power and light-sensing devices to improve our view of the world, and is already being used to enhance the images captured by earth-based telescopes. Here the technique is being employed to get a clearer view of the back of the retina.

The Irish National Nanofabrication Facility (NNF) was set up at the National Microscience Research Centre (NMRC) in Cork in 1999 with funding of €12.7 million from the Higher Education Authority of Ireland. It is the only such facility in Ireland and Britain and allows university and industrial researchers access to an R&D platform. Nanotechnology research at NMRC covers the design, synthesis, fabrication and characterisation of nanostructures and nanosystems. The NMRC objectives are to

- develop a new understanding of nanoscale phenomena and construct new nanoscale structures, devices and systems;
- use these new nanoscale systems as a tool kit to develop new applications in science and engineering.

NMRC aims to provide a complete nanotechnology development loop to enable innovative exploitation of nanosystems specifically within emerging ICT application areas, e.g. nanoscale electronics, and at the interface between ICT and other disciplines, e.g. with photonics (nanophotonics) and with life sciences (nanobiotechnology). One of the SFI-funded projects at the NMRC looks at photonic software and examines the ways to improve the fundamental understanding of photonic materials and devices and enable the design of structures for new capabilities and applications.

The Irish Nanotechnology Association was established in 2002 by Enterprise Ireland to encourage the development of nanomaterials and processes by Irish industry. The key objectives are to

- make companies aware of the benefits of nanomaterials;
- highlight state-of-the-art research ongoing in Ireland and promote technology transfer from academia to industry;

- encourage the development of nanotech companies through spin-offs from the universities and the institutes of technology;
- encourage collaboration between researchers and industry.

The association is managed by the Materials Ireland Polymer Research Centre, a programme in advanced technologies (PAT). Visit www.nanotechireland.com/.

3.4.8 Italy

Italian research is excellent in some fields leading to successes in traditional sectors and in those with medium or high technology content, such as instrument mechanics, robotics, microelectronics, optoelectronics and biomedical technologies. Italy's science and technology guidelines include priority areas for nanotechnology, intelligent materials and sustainable development and climate change and governance in a knowledge-based society. Nanotechnologies and material development are seen as the key to development of other macro-areas such as instrument mechanics, telecommunications, energy, environment, transports, agro-food, health and cultural heritage. With this is mind, there is a strong emphasis on multisectoral enabling technologies.

The automotive sector is actively taking up microsystems and nanotechnology for improving car safety and for catalysts, paints, and structural and functional materials. In the health sector, nanotechnology is being investigated for longer-term applications in pharmacy on chips, nanoparticles and gene therapy, surfaces for medical implants and tissues and organic silicon interfaces. Minatech is an economic and technological intelligence (ETI) project looking at trends in micro- and nanotechnologies and applications and markets for these technologies.

Italian funding for nanotechnology research almost quadrupled in the period 1997–2000. Funding comes from the Ministry of Scientific Research and the National Institute for Physics of Matter (INFM) and the National Research Council (CNR). In the past, CNR funded a national research programme in nanotechnology (1998–2000), with L8 billion (about £2.5 million pounds, or €4 million) of government funding. This programme focused on three lines:

- nanotechnology and molecular devices for electronics;
- nanomaterials and nanodevices for the biomedical sector;
- nanostructures for other applications.

Participating research groups were located in universities and national research centres of CNR and the National Energy Research organisation ENEA.

INFM has invested €3 million in a new laboratory in southern Italy at the University of Lecce dedicated to nanotechnology. Agilent Technologies Inc. and the University of Lecce have signed a technical cooperation agreement in the field of inorganic and organic photonic technologies and devices for fibre-optic communications. The University of Lecce will provide a new laboratory along with state-of-the-art equipment available in the nanotechnology laboratories, including metallorganic chemical vapour deposition (MOCVD) reactors, electron-beam lithography, nanoprocessing, scanning probes and chemical labs. Technical cooperation will be developed in

- epitaxy of new semiconductor materials for telecommunications,
- new-concept nanostructure lasers for telecommunications,
- organic materials and technologies for infrared photonics,
- spatially resolved characterisation of devices at the nanoscale.

These materials technologies will be researched by a team of scientists from the University of Lecce and from Agilent Technologies. The nanotechnology group at the Engineering Faculty of Lecce was until now supported by the National Institute for the Physics of Matter and by the EU.

A new CNR Research Institute in Photonics and Nanotechnology has been established in Rome, as one of a series of 28 new research institutes. The institute will carry out research on devices for photonics, optoelectronics, electronics, laser sources, new materials and characterisation techniques, nanotechnologies and micro- and nanofabrication.

Italian nanotechnology research is strongly related to biotechnology. The Elba Foundation, chaired by the biophysicist Professor Claudio Nicolini, is probably the most internationally visible. This foundation started in 1994 as a follow-up of the Elba Project, an international collaboration in bioelectronics between Russia and Italy. This and other Italian research in bioelectronics is coordinated in the National Bioelectronics Pole (PNB).

3.4.9 Luxembourg

On 31 May 1999 the government of Luxembourg created a National Research Foundation to distribute R&D funds and develop a national research policy. In June 2001 it published its first activity report. In the first 18 months of operation, it has organised two calls for expressions of interest, which received 50 proposals. From these it has created four research programmes:

- SECOM, on electronic commerce security, for €7.5 million;
- NANO, on innovative materials and nanotechnologies, for €6.7 million;
- EAU, on sustainable water resource management, for €5 million;
- SANTE-BIOTECH, on biotechnology and health, for €6 million.

The NANO programme aims to create a European research centre in characterisation of materials in the nanometre range. The materials include plastics, metals, gases, and biological tissues and cells. The Centre will acquire the necessary instruments, including secondary ion mass spectrometry (SIMS), nanomechanical surface analysis, and biocompatible measurement methods.

3.4.10 Netherlands

The aim of the Dutch ministry of Economic Affairs is to increase the innovative capacity of the Dutch economy. It is investing about €5 million extra in innovation in the coming 10 years. Areas for innovation include microsystems technology and nanotechnology. However, until 2000 there was no national research programme dedicated to nanotechnology in the Netherlands. Nanotechnology research was funded through the normal university budgets, from the national research funding organisations NWO, STW and FOM.

The Foundation for Fundamental Research of Matter (FOM) funds a number of research programmes related to nanotechnology. The programme on nanotechnology and nanoelectronics runs from 1998 to 2005 and has a budget of €5.5 million. The programme on nanostructured optoelectronic materials runs from 1999 until 2003 with a budget of €5.8 million. The programme on single-molecule detection and nano-optics runs from 1999 until 2004 with a budget of €2.4 million.

Several interdisciplinary research centres are active in nanotechnology. The most important are Biomade, DIMES and MESA+. Biomade is a commercial centre of excellence in molecular (or bio) nanotechnology, related to the University of Groningen's Biotechnology and Biomedical Research Institute. Opened on 1 January 2000 with a budget of €12 million, it functions as an incubator for start-up companies. The research is carried out in Biomade, and Applied Nanosystems is responsible for commercialising the patented results. Visit www.biomade.nl/.

The Delft Institute of Microelectronics and Submicrontechnology (DIMES) is one of the Netherlands' leading research centres in nanotechnology and related research. Even though the Dutch government does not have a distinct nanotechnology R&D programme, it does fund DIMES and some other centres of excellence at universities. Research in nanoscale electronics and nanoscale structuring is carried out in the Laboratory for Nanoscale Experiments and Technology (NEXT). DIMES' annual budget is around €16 million (£10 million). DIMES also collaborates with established companies, including Akzo-Nobel (plastic solar cells), Leica (e-beam lithography), OptEm and Magma (submicron modelling and extraction). The institute is experienced in European collaborative R&D projects. At present, it leads the MOSIS project on micro-optical silicon systems. Visit http://guernsey.et.tudelft.nl/mosis/.

The MESA+ research institute at the University of Twente specialises in microtechnology and materials. It emerged in mid 1999 after a regrouping of the university's research in microtechnology and materials. Currently 400 people work there and it has a budget of €20 million, 50% of which should come from outside contracts. The Technical University of Eindhoven is installing a new Centre of Expertise for Nano Devices and Materials Design, in which the university itself is investing €24 million in the coming 5–10 years. The research builds on existing expertise in polymers, catalysis and photonics. Wageningen University and Research Institute, the national knowledge centre in agricultural research, is actively promoting a collaboration with Biomade and MESA+ in nanotechnology for agro-applications.

Dreamstart is a government-funded support organisation that aims to organise and develop a technostarter and venture capital market in four high-tech fields, including nanotechnology. It provides an action plan for nanotechnology, claiming a total government investment in nanotechnology research and fostering of spin-offs worth about €107 million in the coming decade. Dreamstart was launched in an effort to couple entrepreneurialism with the Netherlands' proven success as a high-tech business location. During 2003, Dreamstart implemented a promotional campaign targeting would-be entrepreneurs in the fields of nanotechnology, medical technology and food technology. Its mission is to improve the quantity and quality of technology-based start-ups in the Netherlands. In 2002 Dreamstart began founding start-up incubator facilities in conjunction with partner universities.

In 2000 a number of small companies active in development and commercialisation of micro- and nanotechnologies formed MINAC, the Micro and Nano Cluster. The goal of MINAC is to enlarge the knowledge and potency of micro- and nanotechnology for the Netherlands, and for its members in particular, by enabling its members to join forces.

3.4.11 Poland

KBN, the State Committee for Scientific Research, is a governmental body that was set up by the Polish parliament on 12 January 1991. It is the supreme authority on state policy in the area of science and technology. The government's new economic strategy clearly defines the preferred fields of scientific research and development. In accordance with the strategy's plan, here are the priority areas:

- biotechnology including genetic engineering,
- informatics and telecommunication,
- microelectronics and nanotechnologies,
- robotisation and automation,
- new material technologies.

The European Commission has identified a number of academic centres of excellence in Poland. Some of these include expertise relevant for nanotechnology. The Centre of Molecular and Macromolecular Studies of the Polish Academy of Sciences (PAS) in Lodz employs about 150 researchers focusing on structural studies of materials on molecular, macromolecular and supramolecular levels (www.cbmm.lodz.pl).

The Institute of Physics of PAS in Warsaw hosts CELDIS, the Centre for Physics and Fabrication of Low-Dimensional Structures. This consists of 25 laboratories, and focuses on education and research in solid-state physics, mainly semiconductors and magnetic materials. The research is targeted to low-dimensional nanometre structures (http://info.ifpan.edu.pl).

The high-pressure research centre UNIPRESS of PAS has world-class high-pressure equipment for multidisciplinary research, including nanomaterials.

(www.unipress.waw.pl). The Institute of Fundamental Technological Research of PAS hosts a Centre of Excellence in Advanced Materials and Structures (www.ippt.gov.pl/amas). The Institute of Biochemistry and Biophysics of PAS in Warsaw focuses on molecular biology education and research (www.ibb.waw.pl).

Poland hosts two specialised nanotechnology research networks. FAMA unites 21 Polish partners in research in advanced functional materials. UNIPRESS leads the international network in interfacial effects of nanostructured materials, involving 17 partners.

3.4.12 Russia

Support for nanoparticles and nanostructured materials research in Russia and other countries of the former Soviet Union (FSU) dates back to the mid 1970s. The first public paper concerning the special properties of nanostructures was published in Russia in 1976. In 1979 the council of the Russian Academy of Sciences (RAS) created a section on ultra-dispersed systems. Now research strengths are in the areas of preparation processes of nanostructured materials, metallurgical research for special metals and research for nanodevices. Due to funding limitations, characterisation and utilisation of nanoparticles and nanostructured materials requiring costly equipment are less advanced than processing techniques.

In 2000 the Russian Federation spent $850 million on research. In Russia the spending per researcher is only 4% or 5% of the amount spent in the US and Japan. The US spends 26 times as much on research than Russia, Japan 10 times as much. The Russian government and international organisations are the primary research sponsors for nanotechnology in Russia. However, laboratories and companies privatised in the past few years, such as the Delta Research Institute in Moscow, are under development. With a relatively lower base in characterisation and advanced computing, the research focus is on advanced processing and continuum modelling. Research strengths are in the fields of physico-chemistry; nanostructured materials; nanoparticle generation and processing methods; applications for hard materials, purification and the oil industry; and biologically active systems.

Nanotechnology was funded during 1992–2004 under the Physics of Solid State Nanostructures programme. The associated annual conference series, Nanostructures: Physics and Technologies, is a prestigious one in Russia. Eleven RAS institutes and four universities are engaged in 13 nanoprojects in this programme. Funding for nanotechnology is channelled via the Ministry of Science and Technology, the Russian Foundation for Fundamental Research, the Academy of Sciences, the Ministry of Higher Education, and other ministries with specific targets. The Ministry of Science and Technology funds two national programmes in nanoscale science: one on surface science and one on nanochemistry. It is presently developing a new programme in biology.

Since the 1990s, the EU and Russia have collaborated in R&D through the INCO/Copernicus 2 horizontal programme for international cooperation in FP5 and INTAS (www.intas.be) for fundamental research. Russian scientists from 14 institutes are also engaged in international European COST networks (www.belspo.be/cost); Russian companies collaborate in 31 projects in EUREKA (www.eureka.be). Since 1994 small innovative companies have been supported through the Russian Federal Foundation, which assists in the development of small innovative enterprises (http://mch5.chem.msu.su/fond/welcome1.html). So far it has supported 400 proposals involving 60 000 people in 500 enterprises.

3.4.13 Spain

There were no specific nanotechnology funding programmes until 2002, when the funding priority Nanotechnologies, Microtechnologies and Integrated Development of Materials was announced as one of 11 strategic research programmes by the Spanish government. The research plan provides for a large research infrastructure for a multi-purpose X-ray ultrabrilliant pulsed laser. Funding for research comes from

- the Special Supplementary Fund for Research (FISR),
- the Fund for Technological Innovation (FIT–Ministry of Industry),
- the Fund for Investments in Basic Research (FIRB).

Non-targeted basic research programme in the current scientific strategy include

- particle physics and large accelerators,
- construction of the Spanish line (SPLINE) in the ESRF (already begun),
- common elements of the ATLAS and CMS detectors in CERN (already begun).

3.4.14 Sweden

Sweden has one of the highest R&D per capita expenditures in the world, so it is not surprising that its nanotechnology activities have attracted attention. Nanotechnology was taken up relatively early when more than 10 years ago, a research programme was initiated at the analytical chemistry department of the Royal Institute of Technology dealing with concepts related to the nanoscale. Similarly, nanostructure materials research has a long history.

Since 1 January 2001 the Swedish government has restructured its research system. The General University Fund supports 47% of the total current cost of R&D in universities. The Swedish Research Council is responsible for funding basic research in three areas, including natural and engineering sciences. The state and the business sector collaborate through co-funded industrial research institutes. The Knut and Alice Wallenberg Foundation is a private fund that supports expensive scientific equipment and major scientific programmes. (http://wallenberg.org). Vinnova is the national innovation support organisation (www.vinnova.se). There are two major programmes: the Interdisciplinary Materials Research Programme and the Nanochemistry Programme.

The Swedish Foundation for Strategic Research (SSF) is funding a five-year research programme on nanochemistry, which started in 1999. SSF funding amounts to SKr 40 million (€4 million) over the five-year period. The programme aims to develop innovative tools, technologies and methodologies for chemical synthesis, analysis and biochemical diagnostics, in nanolitre to femtolitre domains. It is open to scientific as well as industrial collaborations from inside Sweden and the EU.

The foundation also supports a programme at Chalmers University of Technology on quantum devices and nanoscience, and a five-year research programme on nanochemistry at the Division of Analytical Chemistry of the Royal Institute of Technology in Stockholm. Furthermore, the Swedish research councils support individual projects in nanoscience. The Natural Science Research Council funds nanostructural materials research, which received €0.8 million in 1998; other materials research topics received €4 million. The Swedish Research Council for Engineering Sciences funds similar topics at a level equivalent to about €6.9 million in 1998. SSF funded ten-year consortia which ended in 2000. These consortia are being replaced by consortia related to nanotechnology Table (3.1).

Table 3.1 SSF consotia related to nanotechnology

University	Topic
Linköping University	Biomimetic materials science
Uppsala University	Quantum materials
Uppsala University	Biomimetic enzyme catalysis
Lund University	Advanced molecular materials
Linköping University	Low-temperature thin film synthesis
Royal Institute of Technology	Functional ceramics for sensors and IT
Chalmers University of Technology	Complex oxide materials for advanced devices
Uppsala University	Fundamental research and applications of magnetism
Linköping University	Quantum wires and dots for optoelectronics
Chalmers University of Technology	Carbon-based nanostructures for semiconducting electronics

The Acreo Institute, owned jointly by an industrial association and a state-owned holding company, was created in May 1999, based on the former Institute for Industrial Microelectronics and Institute of Optical Research. Acreo has centres in Kista (north of Stockholm), Norrköping and Lund. Its mission is to promote cooperation between the research world and industry so that research results can be developed and quickly transferred into products and processes for commercial use. Between 1997 and 2000, Acreo and the Royal Institute of Technology developed the world's smallest blood pressure sensor. The sensor is now produced by the Swedish microelectromechanical systems manufacturer Silex, a spin-off from Acreo (www.acreo.se/).

With $80 million in funding, MC2, the Microtechnology Centre at Chalmers University of Technology in Goteborg, Sweden, is leading the way in the Scandinavian country's small-tech research. Funding came from the Swedish government and private institutions, including the Knut and Alice Wallenberg Foundation. Chalmers University of Technology also invested in MC2 (www.mc2.chalmers.se). The varied projects blend micro-, bio- and nanotechnology and form one of the largest groups under the heading of 'microtechnology' at any university in the world. Projects at MC2 range from microwave electronics to microelectronics and nanotechnology. Here are some examples:

- Research on single-electron transistors, which will lead the way to quantum computing, where millions of calculations can be made in a single step. Work is under way on making the first primitive building blocks of a quantum computer.
- Working on ways to measure the conductivity of DNA molecules, in the field of bioelectronics. Research involves studying how molecules and biological specimens can be connected electrically. The group has already produced single-electron transistors made of molecules in the laboratory. This could lead to using the self-assembling properties of molecules to build electronic circuits or even computers.
- Bionics research on nanobiotechnology and communication interfaces between micro- or nanoectronics and biological matter, including living cells. The goal is to communicate with collections of living neurons. This means using brain signals, or 'thought power', to control electromechanical and electro-optical devices, and to make a brain understand signals from external sensors.

3.4.15 Switzerland

Switzerland has a well-integrated innovation system including both federal and cantonal government support. Industries, academia and private R&D have been focusing on micro- and nanotechnology since the mid 1990s. The federal Ministry of the Interior is responsible for science and technology, which is coordinated by the board of the Swiss Federal Institute of Technology, ETH Zürich. Switzerland makes more than SFr 40 million available each year for research in nanotechnology and related fields. Thus on a per capita basis, Switzerland's commitment to nanotechnology is the highest in the world.

The MINAST programme ran from 1996 to 1999, involved SFr 55.6 million of funding from the federal funding council and ETH Zürich and SFr 73 million from others. National research programme NFP36 in nanosciences ran from 1995 until 1999 and involved 40 research groups. It focused on the study and manipulation of mesoscopic and molecular systems on a local scale, mainly through scanning probe methods. Even though Switzerland is not an EU member state, Swiss research groups are frequent members of EU-funded RTD projects, also in nanotechnology. Switzerland is also a member of the COST intergovernmental cooperation in scientific research. EPFL hosts the secretariat of COST action 523 in nanostructured materials. For the funding period 2004–2007 ETH Zürich is coordinating an initiative for micro- and nanotechnology worth about SFr 40 million. This is thought to be sufficient to stimulate further industrial development of nanotechnology.

However, it comes at the end of a concerted programme of funding in the area of nanotechnology.

The main nanotechnology programme in Switzerland is TopNano21, which ran from January 2000 until the end of 2003, and for more fundamental research in supramolecular materials there is national research programme NFP47 (www.snf.ch/NFP/NFP47/home_e.html).

TopNano21 aims to develop the knowledge infrastructure to allow domestic manufacturers to commercialise nanotechnology. The budget for the years 2000 to 2003 was SFr 62 million. The idea is to support entrepreneurs with marketable ideas and it is designed to nurture one start-up a month.

The regions Lausanne–Geneva, Neuchâtel and Zurich are particularly active in micro- and nanotechnology. Leading research institutes with an international standing are CSEM and the Paul Scherrer Institute (PSI). EPFL in Lausanne and ETH Zürich are the federal technological institutes and have a strong nanotechnology research programme. The universities of Fribourg, Basel and Neuchatel collaborate, with PSI and the instrument-manufacturing SMEs CSEM and EMPA, which specialise in organics, ceramics and composites.

ETH Zürich has strong research areas in nanotechnology. The Quantum Photonics Institute specialises in quantum wires and dots for applications in quantum wire light-emitting diodes for optical telecommunication and it has a spin-out company, BeamExpress (www.beamexpress.com). There is a research group focusing on basic properties such as electrical conductance as well as the mechanical strength of carbon nanotubes. Long-term applications are in new IC chips. The Centre for Micro and Nanotechnology at the Engineering Science school focuses on handling nanopowders as well as characterising nanostructured surfaces.

The University of Basel (www.nanoscience.unibas.ch) is the oldest university in Switzerland. The Institute of Physics specialises on research into image detection, scanning probes, optics and nuclear physics. It is the home of the scanning tunnelling microscope, invented by Nobel laureate Gerd Binnig. The National Competence Network, Nanoscale Science, is coordinated at the University of Basel and has eight public and private partners. It runs from 2001 to 2011.

PSI is a multidisciplinary natural science and technology research institute created in 1988 when the Swiss Institute of Nuclear Research was merged with the Swiss Federal Institute of Reactor Research. It provides a user lab for the international research community in universities and industry, offering capabilities in basic and applied research. Its core competencies are solid-state physics, materials sciences, particle physics and astrophysics, life sciences and nuclear and non-nuclear energy research, proton therapy, and micro- and nanotech research.

The Center Swiss for Electronics and Microtechnology, Inc. (CSEM) is a private non-profit organisation with 70 shareholders providing access to technologies. It is supported with long-term contracts provided by the Swiss government designed to finance applied research. Interested in the future, it looks at new markets and high-risk technology, generating $100 million with support of spin-offs, and solving industrial problems through moving basic research to the applied research stage, which bridges a gap in commercial product development.

CERN, the European Laboratory for Particle Physics, is the world's largest particle physics research centre. Founded in 1954, it was one of Europe's first joint ventures and has become a shining example of international collaboration. Some 7000 scientists, half the world's particle physicists, use CERN's facilities. EMPA, the Swiss Federal Laboratories for Materials Testing and Research. consists of a group of laboratories set up in 1880 for construction materials testing, but not greatly involved in research. EMPA is now one of the leading research institutions involved in sustainable materials and material systems technologies. (www.empa.ch). Research in supramolecular materials is funded through national research programme NFP47 running until 2004 with a budget of €10 million over five years.

3.4.16 United Kingdom

The UK had an early interest in nanotechnology, with a DTI National Initiative on Nanotechnology (NION) announced in 1986, followed in 1998 by a four-year LINK nanotechnology programme. The final tranche of funding for LINK projects was handed over in 1996. After this time there was no national strategy for nanotechnology in the UK, although dispersed research involving nanoscale science continued to be funded. In 1997 the Institute of Nanotechnology was created to fill the gap and to act as a focus of interest in nanotechnology. In 2000 the government White Paper 'Excellence and Opportunity–A Science and Innovation Policy for the 21st Century' identified nanotechnology as a vital new and innovative area capable of creating new products and new industries. It recognised that the UK economy could be strengthened by focusing on the research, development and production of biomedical products based on advances in nanotechnology.

In June 2001 the DTI appointed a nanotechnology expert panel to provide guidance on nanotechnology policy in the UK. The first action of the panel was to call for a benchmarking study to be undertaken, to include scenario planning of possible nanotechnology developments in relation to the UK industrial base over the next five years. Nanotechnology was identified as one of the priority areas in the £41 million Basic Technologies programme to provide funding for high-risk research that may result in some new disruptive technological development. In addition to the £41 million for research, the government has also introduced a new £25 million programme over three years aimed at helping businesses commercialise the key technologies emerging from the Basic Technologies programme. Nanotechnology is also one of the four key research priorities in the third round of Foresight Link Awards. These awards have a budget of £15 million.

Some £18 million of ring-fenced funds over six years, for two interdisciplinary research collaborations (IRCs) in nanotechnology, were awarded in 2001 to consortia headed by Oxford and Cambridge Universities. Funds for these collaborations are being made available by three of the government's research councils (EPSRC, BBSRC and MRC) plus the Ministry of Defence. It is the UK government's largest commitment to nanotechnology to date. After six years, the IRCs will revert to conventional means of support. The IRC in nanobiotechnology is headed by Oxford University with the Universities of Glasgow and York, and the Medical Research Council. This collaboration also involves links with the Universities of Cambridge, Nottingham and Southampton.

The main research themes of the Oxford consortium are

- molecular machines,
- functional membrane proteins,
- bionanoelectronics and photonics,
- single-molecule experimental techniques.

The Cambridge IRC will concentrate on the physics of nanotechnology and will focus on the general theme of fabrication and organisation of molecular structures.

Further government funding partly supports a high-technology cluster development initiative to build on activity in nanoscale science and technology at the five universities in north-east England, and includes funding from the private sector and the regional development agency One NorthEast. The regional portfolio includes surface engineering (Northumbria), chemical and biological sensors (Sunderland and Teesside), molecular electronics (Durham) and biomedical nanotechnology (Newcastle). Together with the International Centre for Life in Newcastle, which services the biotechnology sector, the University Innovation Centre for Nanotechnology will act as a cross-sector driver for regional high-technology cluster development.

The Institute of Nanotechnology is the primary source of information on nanotechnology across Europe, and is the lead partner in a £2.7 million European Network Nanoforum. EUSPEN, the European Society for Precision Engineering and Nanotechnology, is based at Cranfield and brings together the academic and industrial community working on ultraprecision engineering. Its aim is to disseminate expertise, future requirements and emerging technologies in ultraprecision engineering, nanometrology and precision metrology, and nanotechnology.

The National Physical Laboratory (NPL) is a world-class centre for nanometrology. It has been involved in nanotechnology since being the coordinator of the UK's National Initiative on Nanotechnology (NION) in 1986. NPL also undertakes a wide range of relevant research relating to many aspects of nanotechnology, including basic science; the production of components with nanofeatures, their characterisation and performance evaluation.

The University of Birmingham has launched I2Nano NanoTech Centre as the first phase of a vision to establish the West Midlands as a global force in the commercial exploitation of nanotechnology. The centre will link industrial companies to the nanoengineering research of the university and its partners, which covers a diverse multidisciplinary portfolio, based on the twin themes of nanoparticles and nanosystems. The venture is supported by the regional development agency Advantage West Midlands, Birmingham City Council and industry as part of a regeneration strategy along the A38 Science Corridor. Current partners have already established nanotechnology groups, including QinetiQ, the Central Laboratory of the Research Councils (CCLRC), AstraZeneca, SKB, Queen Mary College and the DTI. Building on total university nanotechnology grants in excess of £11 million, including participation in a £2.2 million ACORN project, it is anticipated that the centre will be supported through private and public finance.

CCLRC at Daresbury runs an X-ray photoelectron spectroscopy (XPS) service for UK university users to carry out surface analysis related to a wide range of materials science research programmes. CCLRC at the Rutherford Appleton Laboratory (RAL) has micro- and nanofabrication facilities housed in more than 1200 m^2 of class 10 and class 300 clean rooms in its Central Microstructure Facility (CMF). CMF also has a wide range of state-of-the-art equipment for high-resolution pattern definition and pattern transfer, including electron and optical lithography, deposition and plasma dry etching systems. Nanostructure metrology is routinely carried out using a field emission SEM (Hitachi S4000) at the facility.

4

The Vision and Strategy of the US National Nanotechnology Initiative

M. C. Roco
US National Science Foundation

4.1 Motivation

Nanotechnology has opened an era of scientific convergence and technological integration with the promise of broad societal implications. The National Nanotechnology Initiative (NNI) is a long-term visionary program announced in January 2000 that coordinates 22 departments and independent agencies with a total budget of $961 million in fiscal year 2004. This chapter briefly outlines the motivation of this major investment and the key transforming strategies for its implementation. As government investments worldwide approach $4 billion, expectations of nanotechnology commercialization and other potential benefits are raised, and concerns about unexpected societal implications need to be answered to the public's satisfaction. Nanotechnology has evolved into a field of broad international interest, increasing collaboration and stimulating competition.

We know more about single atoms and molecules at one end, and on the bulk behavior of materials and systems at the other end. We know less about the intermediate length scale, the nanoscale, which is the natural threshold where all living systems and man-made systems work. This is the scale where the first level of organization of molecules and atoms into nanocrystals, nanotubes, nanobiomotors, etc., is established. The

Nanotechnology: Global Strategies, Industry Trends and Applications Edited by J. Schulte
 ISBN: 0-470-85400-6 (HB)

basic properties and functions of material structures and systems are defined here and, even more importantly, can be changed as a function of the organization of matter via atomistic and/or 'weak' molecular interactions (such as hydrogen bonds, electrostatic dipoles, van der Waals forces, various surface forces, electrofluidic forces, and DNA assembling). This intermediate length scale, where the role of individual atoms and molecules is measurable and where electrons change their behavior from particle to wave, is increasingly pervasive. At this scale, there is a unity in the scientific treatment in various disciplines and for various areas of relevance, and we are searching for unifying concepts. The intellectual drive towards smaller dimensions, which was essentially enhanced by the discovery of size-dependent novel properties and phenomena, has been the initial drive towards nanoscience. Only since 1981 have we been able to measure the size of an atom cluster on a surface (IBM, Zurich). Ten years later, in 1991, we were able to move atoms on surfaces (IBM, Almaden). After other ten years, in 2002, we assembled the molecules by physically positioning the component atoms. We are still at the beginning of this fertile road of scientific breakthroughs, and nanotechnology is extending the frontiers of knowledge.

The second main reason why nanotechnology has received increased attention in the past few years is the promise of significant societal implications, which include better understanding of nature, efficient manufacturing techniques for almost every human-made object and a new world of products beyond what has been possible with other technologies, molecular medicine, and sustainable development with a cleaner environment and efficient energy conversion. In 1999 we projected that $1 trillion in products worldwide will be affected by nanotechnology by 2015; that would need 2 million nanotechnology workers (Roco and Bainbridge 2001). We have begun not only to see and touch matter at the nanoscale, but also to uncover new phenomena and envision manufacturing processes. The experimental tools and modeling techniques are closing the gap between individual atoms and molecules and the microscale. Once the new behavior has been uncovered at the nanoscale and the ability to manipulate the matter under control at that scale has been tested, the importance of the nanoscale becomes evident.

The promising timeline for the beginning of industrial prototyping and commercialization is the third main reason. The first generation of 'passive nanostructures' applied in coatings, nanoparticles, and bulk materials (nanostructured metals, polymers, and ceramics) have already entered commercial markets, and we are advancing toward systematic design methods. The second generation of 'active nanostructures' such as transistors, amplifiers, targeted drugs and chemicals, actuators, and adaptive structures will be on the same path about five years later (about 2005). The third generation of 'systems of nanosystems' with three-dimensional features, heterogeneous nanocomponents and specific assembly techniques (such as bio-assembly, networking at the nanoscale, hierarchical integration, and new architectures) is estimated to first reach the commercial prototype stage in about 2010. A fourth generation, of 'molecular nanosystems' with heterogeneous molecules, based on macromolecules as nanodevices, biomimetics and new molecular designs is estimated to reach an initial phase of prototype development in about 2020.

Table 4.1 Estimated government nanotechnology R&D expenditures in 1997–2004 (estimation in February 2004 in millions of dollars per year)

Region	1997	1998	1999	2000	2001	2002	2003	2004[c]
W. Europe	126	151	179	200	~225	~400	~650	~900
Japan	120	135	157	245	~465	~720	~810	~900
USA[a]	116	190	255	270	465[b]	697[b]	862[b]	960
Others	70	83	96	110	~380	~550	~800	~900
Total	432	559	687	825	1535	2367	3122	3660
(% of 1997)	(100%)	(129%)	(159%)	(191%)	(355%)	(547%)	(722%)	(847%)

Notes: W. Europe includes countries in the EU-15 and Switzerland; the rate of exchange is $1 = €1.1 until 2002, €0.9 in 2003 and €0.8 in 2004; yen rate of exchange is $1 = ¥120 until 2002, ¥110 in 2003, ¥1.05 in 2004; Others includes Australia, Canada, China, Eastern Europe, Former Soviet Union, Israel, Korea, Singapore, Taiwan and other countries with nanotechnology R&D

[a]A financial year begins in the US on 1 October of the previous calendar year, six months before in most other countries

[b]The actual budget recorded at the end of the respective fiscal year

[c]Preliminary data. Estimates use 'nanotechnology' as defined in Roco *et al.* (2000); this definition does not include MEMS or microtechnology without nanotechnology components

4.2 Government Investment

Table 4.1 tracks nanotechnology government investment around the world since 1997 (Siegel *et al.* 1999) by using the NNI definition of nanotechnology (Roco *et al.* 2000; www.nsf.gov/nano; http://nano.gov). Nanotechnology is working – measuring, manipulating and controlling – at the atomic, molecular and supramolecular levels, at a length scale of approximately 1–100 nm, in order to understand and create materials, devices, and systems with fundamentally new properties and functions because of their small structure. The NNI definition encourages contributions that were not possible before: exploiting specific phenomena, properties, and functions at the nanoscale, which are nonscalable outside of the nanometer domain; and an ability to change those properties and functions at the larger length scales by manipulation of matter at the nanoscale. Microelectromechanical systems (MEMS) and other microtechnologies are complementary to nanotechnology and have not been included in the survey except for those with nanoscale components. Virtually all industrialized countries have in development or have established a plan at the national level in recent years. There are good international opportunities for win-win agreements in the precompetitive research areas. The levels of nanotechnology government investment in Europe, Japan, the US, and other countries in 2004 have increased more than eightfold in the six years since 1997.

The White House's National Science and Technology Committee established in October 1998 the Interagency Working Group on Nanoscience, Engineering, and Technology (IWGN) with the role of preparing a national program, and replaced it with the Subcommittee on Nanoscale Science, Engineering and Technology (NSET) in July 2000 in order to implement the NNI (Figure 4.1). Its goals are to facilitate interagency collaboration for nanoscale R&D, establish R&D priorities and budgets, coordinate planning and program

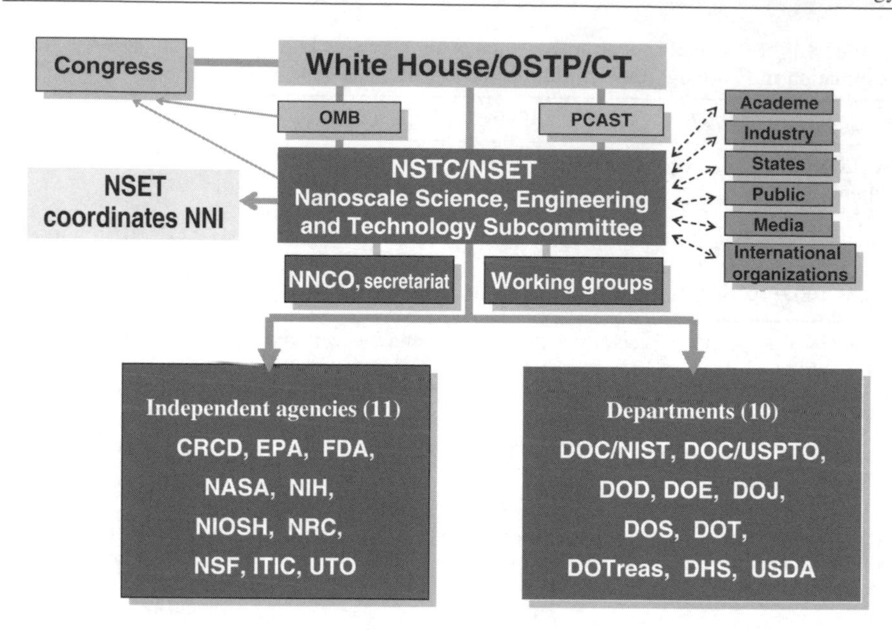

Figure 4.1 Coordination of the US National Nanotechnology Initiative (NNI): OSTP = Office of Science and Technology Policy, NSTC = National Science and Technology Council, CT = Committee on Technology, OMB = Office of Management and Budget, NSET = Nanoscale Science Engineering and Technology (established July 2000), NNCO = National Nanotechnology Coordination Office (established January 2001)

Table 4.2 Contribution of key federal departments and agencies to NNI investment in million dollars per year: each fiscal year (FY) begins on 1 October of the previous year and ends on 30 September

Federal Department or Agency	FY 2000 actual	FY 2001 actual	FY 2002 actual	FY 2003 actual	FY 2004 enacted
National Science Foundation (NSF)	97	150	204	221	254
Department of Defense (DOD)	70	125	224	322	315
Department of Energy (DOE)	58	88	89	134	203
National Institutes of Health (NIH)	32	40	59	78	80
National Institute of Standards and technology (NIST)	8	33	77	64	63
National Aeronautics and Space Administration (NASA)	5	22	35	36	37
Environmental Protection Agency (EPA)	—	6	6	5	5
Homeland Security (TSA)	—	—	2	1	1
Department of Agriculture (USDA)	—	1.5	0	1	1
Department of Justice (DoJ)	—	1.4	1	1	2
Total (% of 2000)	270 (100%)	465 (172%)	697 (258%)	862 (319%)	961 (356%)

implementation, as well as evaluate the NNI to ensure a broad and balanced initiative. The Subcommittee is composed of representatives from agencies and White House officials with interest in the NNI. The Presidential Council of Advisors in Science and Technology (PCAST) provides advice on NNI.

The actual US NNI budget in fiscal year (FY) 2003 was $862 million and the current plan in FY 2004 is $961 million (Table 4.2). The budget decreases in the FY 2004 request noted at NASA and DOD may be explained by the reassignment of applied nanotechnology projects to the respective areas of relevance instead of NNI. State and local organizations committed additional funds for infrastructure, education, and commercialization of more than half of the NNI investment in 2003.

4.3 Transforming Strategy

The main goal of NNI is to fully take advantage of this new technology by a coordinated and timely investment in ideas, people, and tools. NSET has developed a coherent approach for funding the critical areas of nanoscale science and engineering, establishing a balanced and flexible infrastructure, educating and training the necessary workforce, supporting manufacturing and technological innovation leading to potential economic commercialization, promoting partnerships, and avoiding unnecessary duplication of efforts. Here are the key investment strategies.

4.3.1 Focus on Fundamental Research

NNI aims for 'horizontal' interdisciplinary knowledge creation using the same principles, phenomena, tools, and structure architectures in the various areas of relevance. This is combined with 'vertical' transition from basic concepts to technological innovation (Figure 4.2). Nanoscale research is advanced in conjunction with

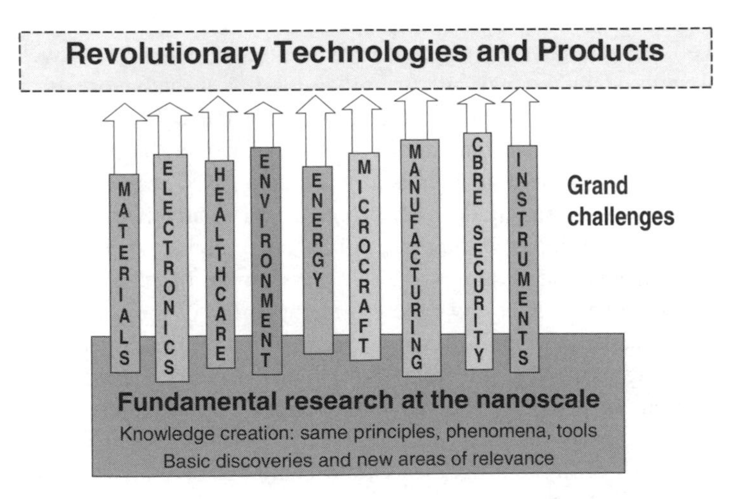

Figure 4.2 Interdisciplinary, or horizontal, knowledge creation versus vertical transition from basic concepts to grand challenges

modern biology, the digital revolution, and cognitive and system sciences, which synergistically support each other and significantly accelerate the overall pace of discoveries and innovations. NNI supports over 2500 active R&D projects in 2004. About 65% of funds go to universities, 25% to government laboratories, and about 10% to the private sector (7% is for small business support on a competitive basis and 3% is for larger organizations).

Nanoscale science and engineering research is intrinsically interdisciplinary and defined by unifying concepts, but it is performed either in an academic environment that rewards primarily individual performance, or in an industrial environment focused on an area of application. Creative approaches are envisioned in order to change the focus from a single discipline to a system approach and from single to multiple areas of relevance.

4.3.2 Long-Term Vision of NNI R&D As Part of a Coherent Science and Technology Strategy

NNI recognizes the importance of visionary, macroscale management measures for the overall success of the initiative. The goals are set for the long term (such as a decade out), and then working backwards, the future R&D needs for each year are identified. The ten-year vision is revisited every five years, with program planning each year, and organizational measures for the implementation of the program each month (Figure 4.3). This is applied by considering various components of the architecture of contributing factors. It includes defining the vision of nanotechnology and NNI goals, establishing the R&D priorities and interagency implementation plan, integrating short-term technological developments into the broader loop of long-term R&D opportunities and societal implications, using peer review for NNI, developing a suitable legal framework, and integrating some international efforts.

First, we have established a vision that is focused more on the novel phenomenological and system behavior and manufacturability at the nanoscale and less on the advantages of smallness itself. Work done under the US National Science and Technology Council (NSTC) has allowed us to effectively address such broader issues. A ‘bio-inspired’ funding approach of the major NNI research areas has been adopted: agencies issue solicitations for proposals for relatively broad R&D themes according to their mission, and then researchers respond with specific ideas in their proposals in a manner suggesting ‘bottom-up’ assembly of each theme.

It is estimated that in 2001–5 we are at the beginning of the fast-rising sector of the classical *S*-shaped development curve where the rate of discovery is increasing. Overall, the long-term NNI research strategy in the interval 2001–5 is balanced across five kinds of activity: fundamental research, grand challenges, centers and networks of excellence, research infrastructure, as well as ethical, legal and social implications and workforce programs. Grand challenges are areas where potential breakthroughs could provide major, broad-based economic benefits, as well as dramatically improve the quality of life.

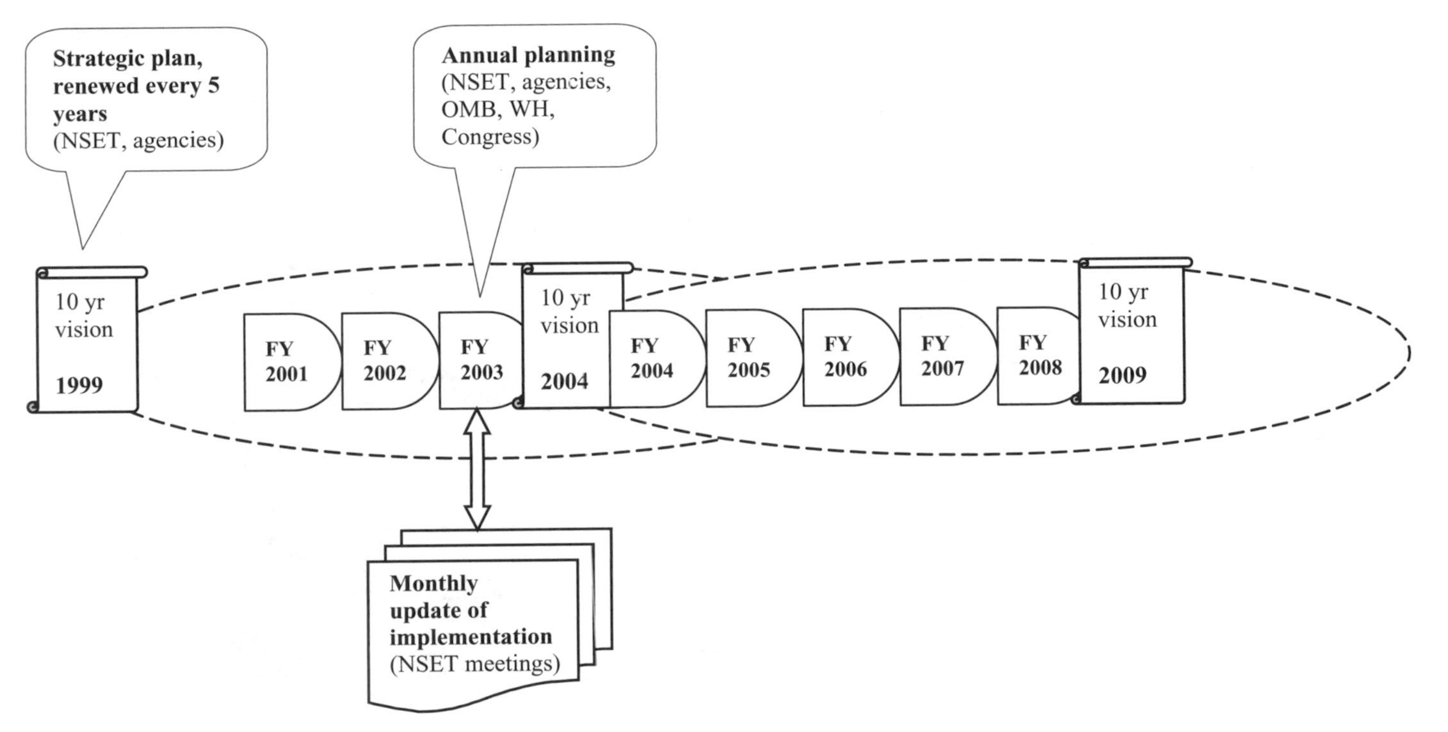

Figure 4.3 Three scales for NNI planning and implementation: 10 years, 1 year and 1 month

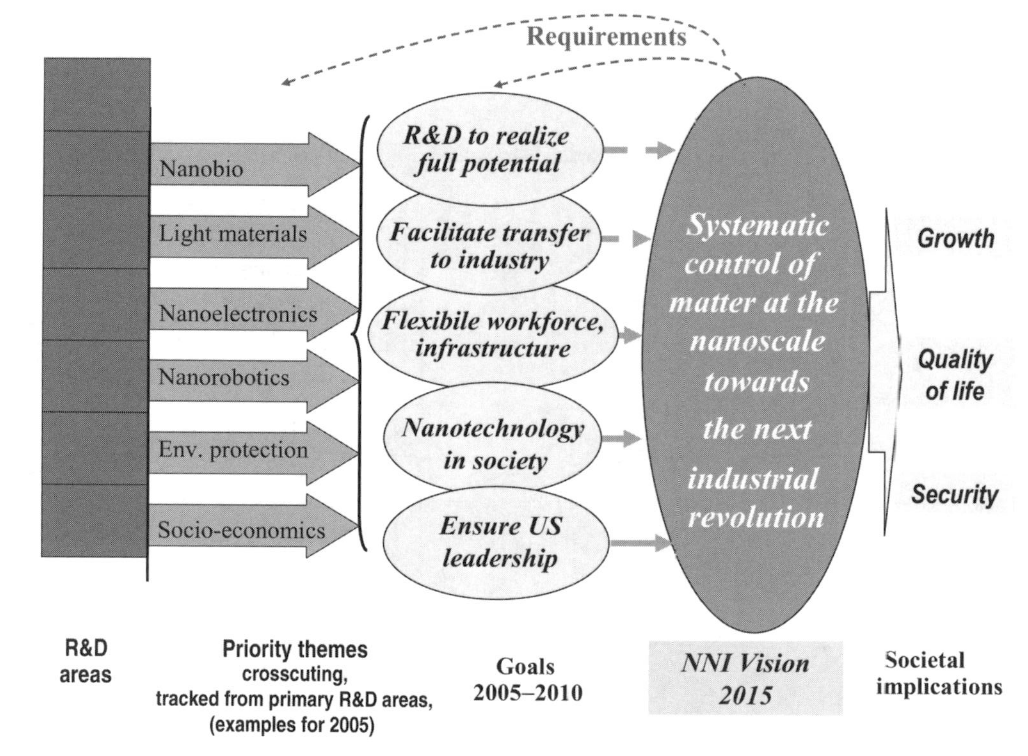

Figure 4.4 Reaching the NNI vision

In FY 2001, NNI identified nine areas of grand challenges (National Science and Technology Council 2000). Nanobiotechnology and nanobiomedical research has progressively increased in importance (National Institutes of Health 2000). In 2002 three new grand challenges were added, related to manufacturing at the nanoscale, instrumentation, and chemico-biological-radioactive-explosive detection and protection. The second strategic planning of NNI has been completed in December 2004 (NSET, 2004) based to the new knowledge and technological foundation developed in the first four years of NNI (Roco 2004). The long-term vision has been established first, and then we have determined the requirements for shorter-term goals and priority themes (Figure 4.4).

Nanoscale manufacturing R&D is an example of a long-term objective of developing systematic methods for economic synthesis and fabrication of three-dimensional nanostructures, establishing nanoscale manufacturing capabilities, and establishing the markets for nanotechnology producers and users. Another important challenge is establishing standardized and reproducible microfabricated approaches to nanocharacterization, nanomanipulation, and nanodevices.

The centers and networks of excellence encourage long-term, system-oriented projects, research networking, and shared academic user facilities. These nanotechnology research centers will play an important role in the development and utilization of specific tools, and in promoting partnerships in the coming years (Tables 4.3 and 4.4).

NSF will run two user networks – the National Nanotechnology Infrastructure Network and the Network for Computational Nanotechnology – and twelve nanoscale science and engineering centers and continue support for thirteen materials research science and engineering centers with research at the nanoscale. DOE has established five large-scale user facilities – the Nanoscale Science Research Centers – NASA four nano-bio-info research centers, DOD three centers, and NIH several visualization and instrumentation centers.

In planning for the future, NNI has been prepared with the same rigor as a scientific project, including a long-term vision developed in 1999 (Roco *et al.* 2000; National Science and Technology Council 1999, 2000; http://nano.gov). The National Research Council (NRC) reviewed NNI in 2002 (National Research Council 2002), and made a series of recommendations such as increasing R&D investment on nanobiosystems and societal implications.

Two bills for nanotechnology submitted in 2003 in the US Congress addressed the need for coherent, multi-year planning with increased interdisciplinarity and interagency coordination. Senate bill S189, 21st Century Nanotechnology R&D Act, in the 108th Congress recommends a five-year National Nanotechnology Program. It was introduced by a group of senators led by Ron Wyden (Democrat, Oregon) and George Allen (Republican, Virginia). The draft bill in the House was HR 766, Nanotechnology Research and Development Act of 2003; it was introduced by a group of representatives led by Sherwood Boehlert (Republican, New York) and Michael Honda (Democrat, California). The two bills were approved by President Bush in December 2003 along with Public Law 108-153. Societal goals

Table 4.3 NNI centers and networks of excellence

	Institution	Year initiated
NSF		
Nanoscale Systems in Information Technologies, NSEC (Nanoscale Science and Engineering Center)	Cornell University	2001
Nanoscience in Biological and Environmental Engineering, NSEC	Rice University	2001
Integrated Nanopatterning and Detection, NSEC	Northwestern University	2001
Electronic Transport in Molecular Nanostructures, NSEC	Columbia University	2001
Nanoscale Systems and Their Device Applications, NSEC	Harvard University	2001
Directed Assembly of Nanostructures, NSEC	Rensselaer Polytechnic Institute	2001
Nanobiotechnology, Science and Technology Center	Cornell University	2000
Integrated and Scalable Nanomanufacturing, NSEC	UC Los Angeles	2003
Nanoscale Chemical, Electrical, and Mechanical Manufacturing Systems, NSEC	UIUC	2003
Integrated Nanomechanical Systems, NSEC	UC Berkeley	2004
High Rate Nanomanufacturing, NSEC	Northeastern University	2004
Affordable Nanoengineering of Polymer Biomedical Devices, NSEC	Ohio State University	2004
Nano-bio Interface, NSEC	University of Pennsylvania	2004
Probing the Nanoscale, NSEC	Stanford University	2004
Templated Synthesis and Assembly at the Nanoscale, NSEC	University of Wisconsin, Madison	2004
DOD		
Institute for Soldier Nanotechnologies	MIT	2002
Center for Nanoscience Innovation for Defense	UC Santa Barbara	2002
Nanoscience Institute	Naval Research Laboratory	2002
NASA		
Institute for Cell Mimetic Space Exploration	UCLA	2002
Institute for Intelligent Bio-Nanomaterials and Structures for Aerospace Vehicles	Texas A&M	2002
Bio-Inspection, Design and Processing of Multi-functional Nanocomposites	Princeton	2002
Institute for Nanoelectronics and Computing	Purdue	2002

and R&D were discussed at each of the previous Congressional nanotechnology hearings, including one on 19 March 2003, and a special hearing on this topic was held on 9 April 2003 by the House Committee on Science. The hearing suggested the need to increase funding in this area and to involve social scientists from the beginning in large NNI projects.

Table 4.4 NNI R&D user facilities

	Institution	Year initiated
NSF		
National Nanotechnology Infrastructure Network (NNIN): a network of 13 academic facilities	Main node at Cornell University	2004
Network for Computational Nanotechnology (NCN): a network of 7 academic facilities	Main node at Purdue University	2004
DOE		
Center for Functional Nanomaterials	Brookhaven National Laboratory	
Center for Integrated Nanotechnologies	SNL and LANL	
Center for Nanophase Materials Sciences	Oak Ridge National Laboratory	
Center for Nanoscale Materials	Argonne National Laboratory	
Molecular Foundry	Lawrence Berkeley National Laboratory	

4.3.3 Policy of Inclusion and Partnerships, Including Promoting Interagency Collaboration

This strategy applies to various disciplines, areas of relevance, research providers and users, technology and societal aspects, and international integration. The vision of a 'grand coalition' of collaborating universities, industry, government laboratories, government agencies, and professional science and engineering communities was proposed in 1999 (Roco *et al.* 2000: V–VIII) and has been implemented through NNI. The added value by synergy in science and technology resulting from partnerships is one of the main reason of establishing NNI. A starting point was the collaborations and monthly working meetings of currently 21 federal agencies covering almost all relevant areas of nanotechnology (Figure 4.5).

Coordination between agencies is a key task of the NSTC's Subcommittee on Nanoscale Science, Engineering and Technology (NSET). It coordinates planning and budgets of the participating agencies, identifies promising research directions, encourages collaborative investments, avoids duplication of effort, and ensures development of a balanced infrastructure. The National Nanotechnology Coordinating Office (NNCO) serves as secretariat to NSET providing technical and administrative support to implement the interagency activities and prepare planning and assessment documents. For example, NSET has coordinated the establishment of new centers and facilities with complementary functions that are being developed by the different agencies.

In addition to industry, an increased role of states and universities in funding nanotechnology has been evident in the US since 2002. Examples are the states of New York (the Albany Nanotechnology Center), California (the California Nanosystems Institute with additional matching from industry at a ratio of 2:1), Illinois (the Institute for Nanotechnology, with joint funding from Northwestern University,

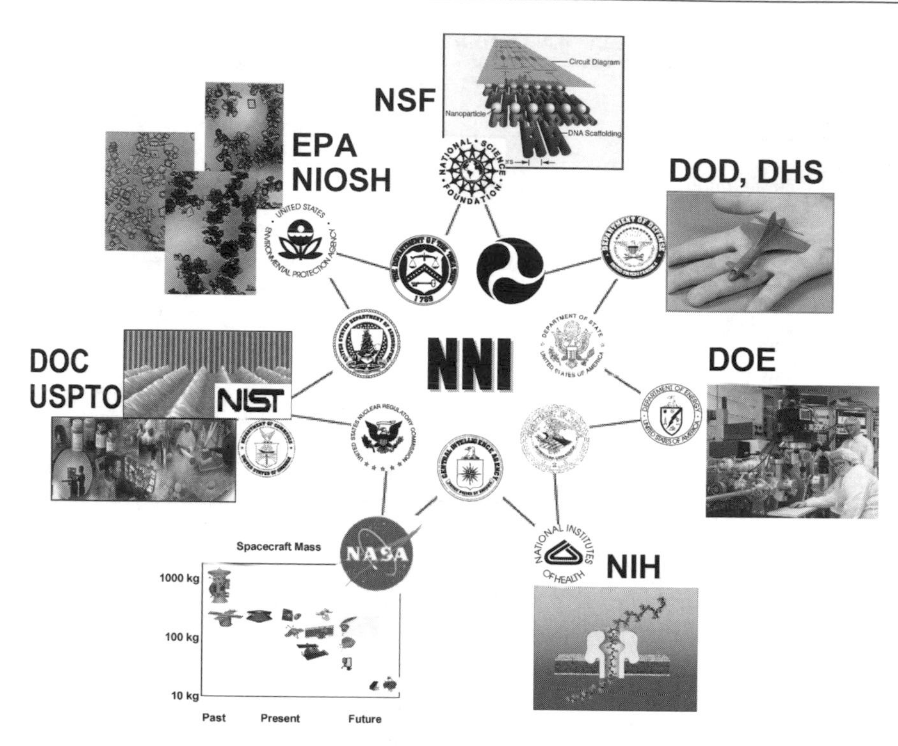

Figure 4.5 NNI embraces 21 federal departments and independent agencies covering various societal needs

and the Center for Nanofabrication and Molecular Self-assembly, with other funding agencies), Pennsylvania (the Franklin Institute for developing partnerships in nanotechnology), Georgia (a new center) and Indiana (contributions to the nanotechnology investment at Purdue University). It is estimated that US industry made about the same level of investment in nanoscale science and engineering research as the federal government in 2003, but it is generally directed to 'vertical' transformations of a fundamental discovery into a product, whereas the federal investment is generally directed to 'horizontal' basic discoveries of relevance to multiple disciplines and areas of relevance. International collaborations are part of the overall partnerships and they are increasing in importance.

4.3.4 Preparation of a Diverse Nanotechnology Workforce

A major challenge is to educate and train a new generation of workers skilled in the multidisciplinary perspectives necessary for rapid progress in nanotechnology. The concepts at the nanoscale (atomic, molecular, and supramolecular levels) should penetrate the education system in the next decade in the same way that microscopic approaches made inroads in the past fifty years. NSF has a plan for systemic and earlier nanoscale science and engineering education. The R&D workforce is managed using merit review and individual incentives. It is estimated that about 2 million nanotechnology workers will be needed worldwide in 10–15 years. One way to ensure a pipeline of new students into the field is to promote interaction with the public at large. Since 2002 several US universities have reported increased numbers of highly qualified students moving into physical and engineering sciences because of the NNI.

Timely education and training will begin moving concepts from the microscopic world to the molecular and supramolecular levels. Changes in teaching from kindergarten to graduate school, as well as continuing education activities for retraining, are envisioned. An important corollary activity is the retraining of teachers themselves. One may consider changes in how we structure information on nanotechnology (Yamaguchi and Komiyama 2001) in order to improve learning and disseminate the results. Five-year goals for NNI include ensuring that 50% of research institutions' faculty and students have access to the full range of nanoscale research facilities, and enabling access to nanoscience and engineering education for students in at least 25% of research universities. Here are three illustrations:

- NSF's Nanotechnology Undergraduate Education program has made about 70 awards in FY 2003 and FY 2004. Nanotechnology grade 7–12 education has been funded through a national center at the Northwestern University and an increased focus on public education is planned in 2005.
- In 2004 a coherent plan has been developed to integrate high-school, technological, undergraduate, and graduate education into a collaborative environment.
- The software NanoKids (Tour 2003) has been developed for interactive learning using video animation on easily accessible computers (Figure 4.6).

Figure 4.6 NanoKids: interactive teaching software for high school. Reproduced with permission from Tour (2003)

4.3.5 Address Broad Societal Goals

The first report on societal implications of nanoscience and nanotechnology (Roco and Bainbridge 2001) was prepared at the onset of NNI in September 2000, and its recommendations were reflected in the NSF program announcements and the operation of NNCO. Nanoscale science and engineering will lead to better understanding of nature, economic prosperity, and improved health, sustainability, and peace. This strategy has strong roots and may bring people and countries together. An integral aspect of NNI's broader goals is increasing productivity by applying innovative nanotechnology for commerce (manufacturing, computing and communications, power systems, energy). Taking this road towards broader goals may bring large benefits in the long term. Aiming at broad societal goals was one of the initial strategies of NNI (Roco 2003), and it has expanded to converging technologies from the nanoscale for improving human performance (Roco and Bainbridge 2003).

Since October 2000 the annual NSF program announcement has included a focus on ethical, legal, and societal implications and on workforce education and training. Research on societal and educational implications will increase in importance as novel nanostructures are discovered, new nanotechnology products and services reach the market, and interdisciplinary research groups are established to study them. The NNI annual investment in research with societal and educational implications in 2004 is estimated at about $45 million (of which NSF awards about $40 million), and in nanoscale research with relevance to environment and health and safety at about $90 million (of which NSF awards about $40 million, NIH about $33 million and EPA about $6 million). The total of about $90 million is approximately 10% of the NNI budget in FY 2004. One example of a supported project is cleaning contaminated soil using iron nanoparticles that are partially coated with other metals (Figure 4.7). This project received joint support from NSF and the Environmental Protection Agency (EPA).

Societal implications include the envisioned benefits from nanotechnology as well as second-order consequences, such as potential risks, disruptive technologies, and ethical aspects. Long-term developments of the field depend on the way one addresses the 'societal challenges' of nanotechnology (Lane 2001). NSET is actively seeking input from research groups, social and economical experts, professional societies, and industry on this issue.

4.4 Closing Remarks

I would like to close this brief overview of NNI with several comments about international collaboration in the future. Nanoscale science and engineering R&D is mostly in a precompetitive phase. International collaboration in fundamental research, long-term technical challenges, metrology, education, and studies on societal implications will play an important role in the affirmation and growth of the field. The US NNI develops in this context. The vision-setting and collaborative model of NNI has received international acceptance.

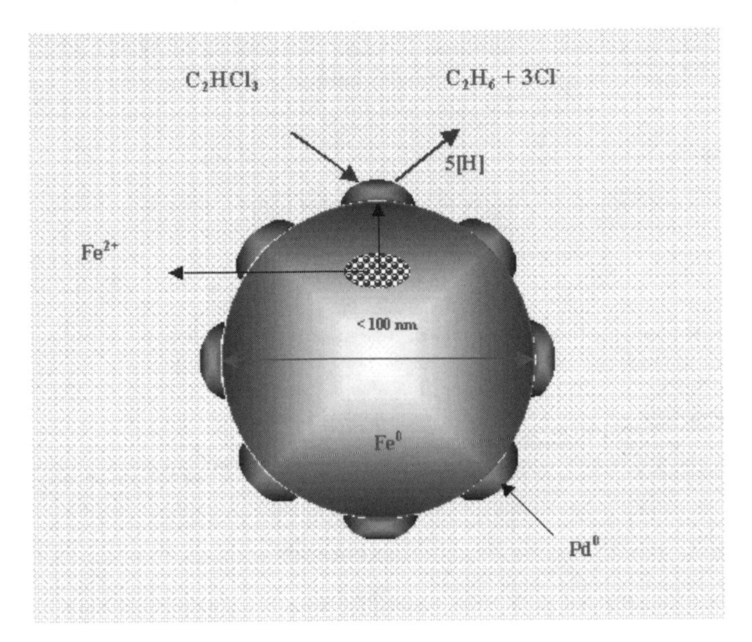

Figure 4.7 Cleaning the environment with iron nanoparticles. Reproduced with permission from Zhang (2003)

Opportunities for collaboration towards an international nanotechnology effort, particularly in the precompetitive areas, will augment the national programs. One may note that large companies rely heavily on R&D results from external sources (about 80% in 2001), of which a large proportion is from other countries (Europe 35%, Japan 33%, US 12%, according to E. Roberts, MIT, at the Sloan School of Management). An increased number of companies are acting globally with a significant flow of ideas, capital, and people. This trend will accelerate and will be the environment in which nanotechnology will develop.

Priority goals may be envisioned for international collaboration in nanoscale research and education: better comprehension of nature, increased productivity, sustainable development, and development of humanity and civilization. Examples include understanding single molecules and the operation of single cells, improving health and human performance, enhancing simulation and measuring methods, creating assembly and fabrication tools for the building blocks of matter, and developing highly efficient solar energy conversion and water desalinization for sustainable development.

Acknowledgements

Opinions expressed here are those of the author and do not necessarily reflect the position of NSET or NSF. This chapter is based on a presentation made at the

National Nanotechnology Initiative Conference, Infocast, Washington, DC, on 3 April 2003; several items were updated before publication.

References

1. Lane, N. Grand challenges of nanotechnology. *Journal of Nanoparticle Research* **3**(2/3), (2001) 1–8.
2. National Institutes of Health. *Nanoscience and Nanotechnology: Shaping Biomedical Research* (2000) NIH, Washington, DC (http://nano.gov or http://grants.nih.gov/grants/becon/becon_funding.htm)
3. National Research Council. *Small Wonders – Endless Frontiers: A Review of the National Nanotechnology Initiative* (2002) National Academies Press, Washington, DC.
4. National Science and Technology Council. *Nanotechnology – Shaping the World Atom by Atom* (1999) Brochure for the public, NSTC, Washington, DC (http://nano.gov).
5. National Science and Technology Council. *National Nanotechnology Initiative: The Initiative and Its Implementation Plan* (2000) NSTC, Washington, DC (http://nano.gov).
6. National Science and Technology Council, National Nanotechnology Initiative Strategic Plan, Dec. 2004, Washington, D.C. (http://nano.gov)
7. Roco, M. C. Broad societal issues of nanotechnology. *Journal of Nanoparticle Research* **5**(3/4), (2003) 181–189.
8. Roco, M. C. The U.S. National Nanotechnology Initiative after 3 years (2001–2003). *Journal of Nanoparticle Research* **6**(1), (2004) 1–10.
9. Roco, M. C. and Bainbridge, W. (eds). *Societal Implications of Nanoscience and Nanotechnology* (2001) NSF and Kluwer, Boston MA.
10. Roco, M. C. and Bainbridge, W. (eds). *Converging Technologies for Improving Human Performance* (2003) Kluwer, Boston MA. First published in June 2002 as an NSF-DOC report.
11. Roco, M. C., Williams, R. S. and Alivisatos, P. (eds). *Nanotechnology Research Directions* (2000) Kluwer, Boston MA. First published in 1999 as an NSTC report.
12. Siegel, R. W., Hu, E. and Roco, M. C. (eds). *Nanostructure Science and Technology* (1999) NSTC and Kluwer, Boston MA.
13. Tour, J. NanoKids. *Seminar presented at National Science Foundation*, (2003) NSF award 0236281, Arlington VA.
14. Yamaguchi, Y. and Komiyama, H. Structuring knowledge project in nanotechnology materials program launched in Japan. *Journal of Nanoparticle Research* **3**(2/3), (2001) 1–5.
15. Zhang, W. Nanoscale iron particles for environmental remediation: an overview. *Journal of Nanoparticle Research* **5**(3/4), (2003) 323–332.

Part Two

Investing in Nanotechnology

5

Growth through Nanotechnology Opportunities and Risks

Jurgen Schulte
Asia Pacific Nanotechnology Forum

It was not the computer box on the desk itself that triggered the tremendous growth in the IT industry. There were computers on our desks for about 20 years before a significant growth in the IT industry was observed. What triggered that growth was a sequence of enabling technologies that made it possible to easily collaborate and communicate beyond the boundaries of the office cubicle. At first, the increasing data density of hard disks made it possible to store applications which became increasingly useful to the general office environment. The advent of computer network technology reduced the cost of communication, and hence collaboration, to an almost negligible amount of a normal business or household operation. Network technology as enabling technology behind our desktop and mobile computers has lifted the computer industry to heights unthought of at the time when the first computer box revolutionized business operation.

While mainstream computer technology is more or less based around a primary electronics industry, we see nanotechnology emerging at the core of many industries. This is easily understood when looking at the nanometer scale, a size scale of importance to all manufacturing and processing industries (pharmaceutical,

Nanotechnology: Global Strategies, Industry Trends and Applications Edited by J. Schulte
 ISBN: 0-470-85400-6 (HB)

electronics, biotechnology, cosmetics, polymers, metal, textile, power, etc.). At this stage, nanotechnology is still at the very beginning of establishing fundamental technologies at the nanoscale, which means at a stage of development that is comparable to the level when the first experiments with vacuum tubes were made that later led to the transmission of radio signals over long distances and other pre-transistor applications of this technology. That is not to say that nanotechnology today is short of producing technologies that have a use beyond research laboratories. There are already many products for general use on the market that have been nanotechnology enhanced, as illustrated in this book. The fundamental differences between the electronics industry and the emerging nanotechnology industry is that development of nanotechnology is driving innovation in many seemingly unrelated industries (e.g. construction, textiles, cosmetics) at the same time, each of them directly producing unique, original products in their own right. This has been quite different in the electronic industry where electronic components were produced which only resulted in another new electronic product, opposed to, for example, an enhanced cotton fibre or strength improved concrete.

Why is it that nanotechnology is turning up in so many different areas at the same time? It is the nature of the nanoscale itself that makes this multi-industry, and increasingly multidisciplinary, development happen. There is a rich body of engineering knowledge at the micrometer scale level (10^{-6} m, macroscopic scale) across all major manufacturing and producing industries. There is an equally good fundamental knowledge at the atomic scale (10^{-10} m, microscopic scale). The phenomena that are observed at these two different size scales are very different. Taking advantage of the unique properties of the respective other size scale may not be of much value to the particular industry or simply not even possible. This is very different at the nanoscale (10^{-9} m, nanometre, mesoscopic scale). Here previously unknown phenomena are being observed that may be turned into useful applications in very many areas. Engineering at the macroscale (10^{-6} m) is relatively inexpensive (a notable exception is the computer chip development) and relatively fast compared to engineering at the molecular or atomic level. On the other hand, at the molecular and atomic level, the number of custom features that one can build is almost unlimited. The development costs, however, are very high and the scaling up of production from a single molecule to a stable bulk production may not be straightforward.

At the nanoscale, which lives between the macroscale and the microscale, it seems that in terms of engineering there is a good chance that all the good things found close to the top (micro) and the bottom of the scale (atom, molecule) can be combined to produce something entirely superior or new. Engineering at this scale is relatively inexpensive and the features of technological interest (and commercial interest) are so much improved and in many cases also find an application is a completely different industry than the underlying materials would have suggested. Traditional mass production techniques and manufacturing processing may need to be adjusted to also cater for production of nano enhanced production. Here self-assembly is probably the most prominent manufacturing and processing technology due the relatively low costs and ease at which it can be scaled up for mass production. In fact, with a minimum of investment and a small set of researchers at hand, one can start a small company literally out of a garage and become a serious supplier, or even competitor, to well-established players in the field.

Materials and equipment costs are relatively low and fabrication methods are relatively easy to learn, and many development processes are similar to already known processes. Technological know-how is only a minor barrier. Processing and manufacturing in nanotechnology has become relatively cheap, which makes prospective profit margins more attractive to small players. The initial investment needed to develop and manufacture at the nanoscale is fairly low compared to other high-tech areas (just as in the old days of emerging desktop computer software development), which makes it possible for the very many bright researchers in less developed countries such as China, India, Malaysia, Thailand, Taiwan and Vietnam to start their own nanotechnology research and even commercial ventures.

Unlike in other emerging industrial revolutions, nanotechnologies are being developed simultaneously in many different industries and rapid cross fertilization of ideas is taking place. This is one of the leading reasons why development in nanotechnology is moving at such a rapid pace. As with all new innovations in technology that move at very fast pace, tangible outcomes can have an incremental enabling, as well as disruptive, effect on business and industry sectors. For instance, a smart invention coming out of the surface chemistry lab can easily move into and eventually dominate the textile industry (incremental innovation) while a soot, carbon black, and nanotube producer (FET displays) may become best friends with the electronic display industry (which already is on the path of wiping out the good old fashion CRT TV box as well as the only recently introduced consumer LCD displays).

Enabling incremental and disruptive stages of Nanotechnology are more clearly illustrated in Table 5.1 at the example of nanostructure engineering.

Although nanotechnology has some disruptive nature, it is possible that the disruptive nature of the (nano) technology itself has been overestimated in the past. New nanotechnology discoveries so far have always been accompanied by an incremental cost recovering, which gave most senior management across industry sufficient time to react. Those, of course, who did not react quickly enough experience new nanotechnology developments to be rather disruptive.

A prominent example of a potentially disruptive nanotechnology has its origin in the discovery of carbon nanotubes, which if braided into a cable are thousands of

Table 5.1 Nanostructure engineering as enabling and disruptive nanotechnology

	Enabling	Disruptive
Nanostructures	Better versions of current devices	Replacing microstructures
Nano-assembly	Enabling nanodevices	Replacing microtechnology processing
Nanotechnology	Enabling nanotechnology through nanoscale toolboxes	Replacing early nanodevices

times stronger that any previous engineered fibre or cable. While nanotubes find their application in the fibre industry, they are now emerging in the electronic display industry as well as in the high-density power battery industry. Spreading across industry has occurred only within the past few years. The pace at which nanotechnology has been taken from the research and development phase to a commercially competitive product platform has made manufacturers and developers realize that their thinking about fundamental innovation cycles may need drastic immediate adjustment. For instance, it took only three years from the discovery of electron emission in carbon nanotubes to the making of super flat, super bright electronic display applications, but it took over 20 years from the discovery of a semiconducting *pn* junction to a transistor consumer application.

Other discoveries like truly self-cleaning, non-stick, highly scratch-resistant surfaces, intelligent paints, etc., went through similar rapid evolution. It has become a new challenge for industry to adjust to the rapid speed of product development based on nanotechnology in order to be able to adopt innovations in nanotechnology at a similarly rapid speed. In some cases it can mean that an entire manufacturing process or product line will no longer be competitive if it is not adjusted in time. It also means that it has become essential for industry to rapidly learn the language of other industries and disciplines so it can assess emerging competitors, sometimes from a completely different field, and to spot developments that can be adopted and those that may dig into one's own market share in the future.

A rapid pace of innovation in nanotechnology does not necessarily mean that economic growth through nanotechnology will come at an equally fast pace, although an initial player with the right product may indeed grow very fast. There are many obstacles to overcome and which are not unlike those that we have seen in the personal computer industry until it was later called the IT revolution. At this stage, we are looking at a rapid spread of interest in nanotechnology throughout a vast landscape of industries. Figure 5.1 illustrates the current landscape of nanotechnology industries that are actively involved in developing nanotechnology. There is more space for other industries to simply adopt nanotechnology developments for value-added existing products (derived products).

Currently, promising technology is rapidly turned into some revenue-generating intermediate product in order to prove its underlying technology concept, and of course to generate funds for further development. Those technologies for which there is no time or resources allocated for further product development or for which there is no immediate idea for a revenue-generating application, are finding a place in the company patent portfolio for potential future use. Other companies, such as Hewlett-Packard, pursue a different strategy. Instead of aiming for a product enhancement or an entirely new nanotechnology-based consumer product line, Hewlett-Packard is developing a strategic patent portfolio that aims at future nanotechnology-enabling nanotechnology i.e. technology that makes it possible to make use of nanotechnology or to build new nanotechnology i.e., the kind of technology that is as fundamental and necessary to the functioning of, say, electronic nanodevices as the copper wire in our office wall for the functioning of our entire office.

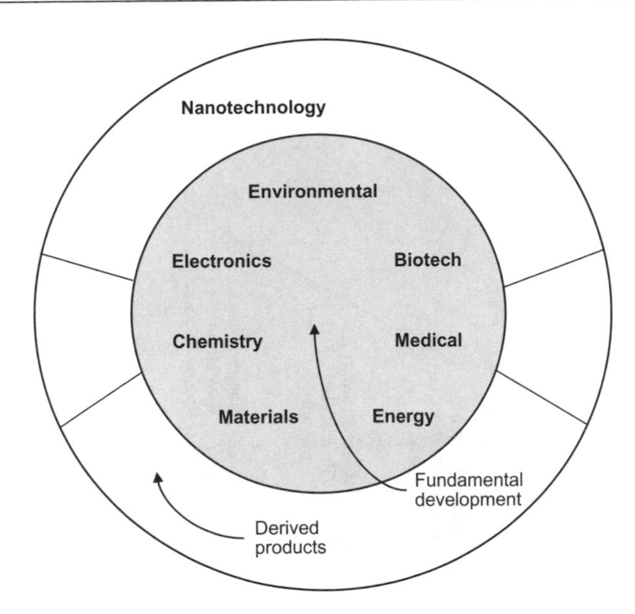

Figure 5.1 Cross-fertilization may bring a wave of new and very fundamental developments, causing more industries to adopt ready-to-use nanotechnology components and tools

Real economic growth through nanotechnology can be expected after a number of serious obstacles have been overcome. One major obstacle is the acceptance or widespread adoption of certain key underlying nanotechnology processes or products. A key underlying technology is one that has been proven to work reliably, is cost-effective, and is generally seen to be there for a long time, such as a selective welding or glue technology with nanoprecision, or a standardized surfaces treatment with a toolbox for adding functionalizations. Along with the emergence of such key technologies, another major obstacle has to be overcome, the industry standards and regulations that are yet to be established.

It is very convenient to be able to travel around the world with a notebook computer and plug in the RJ45 jack to connect to the internet. Similar convenience needs to be set in place for industry sectors to buy nanotechnology raw materials in bulk or nanoscale 'gadgets' for making a new product or for adding features to an existing product but with minimal change to the production line or existing safety requirements. A third major obstacle to be overcome is finding key technologies for nanotechnology-enabled nanotechnology. If there is a nanotoolbox available with a set of tools that can make, combine, adjust or assemble existing nanotechnology, a multitude of products can be manufactured for applications that we cannot imagine today. Nanotechnology will then be ready to be successfully incorporated into almost all existing products from manufacturing tools to ordinary consumer items. That is when nanotechnology will become a major driver of the global economy.

Figure 5.2 illustrates the ingredients and technologies that are driving nanotechnology in a wide range of industries. During its evolutionary development from a

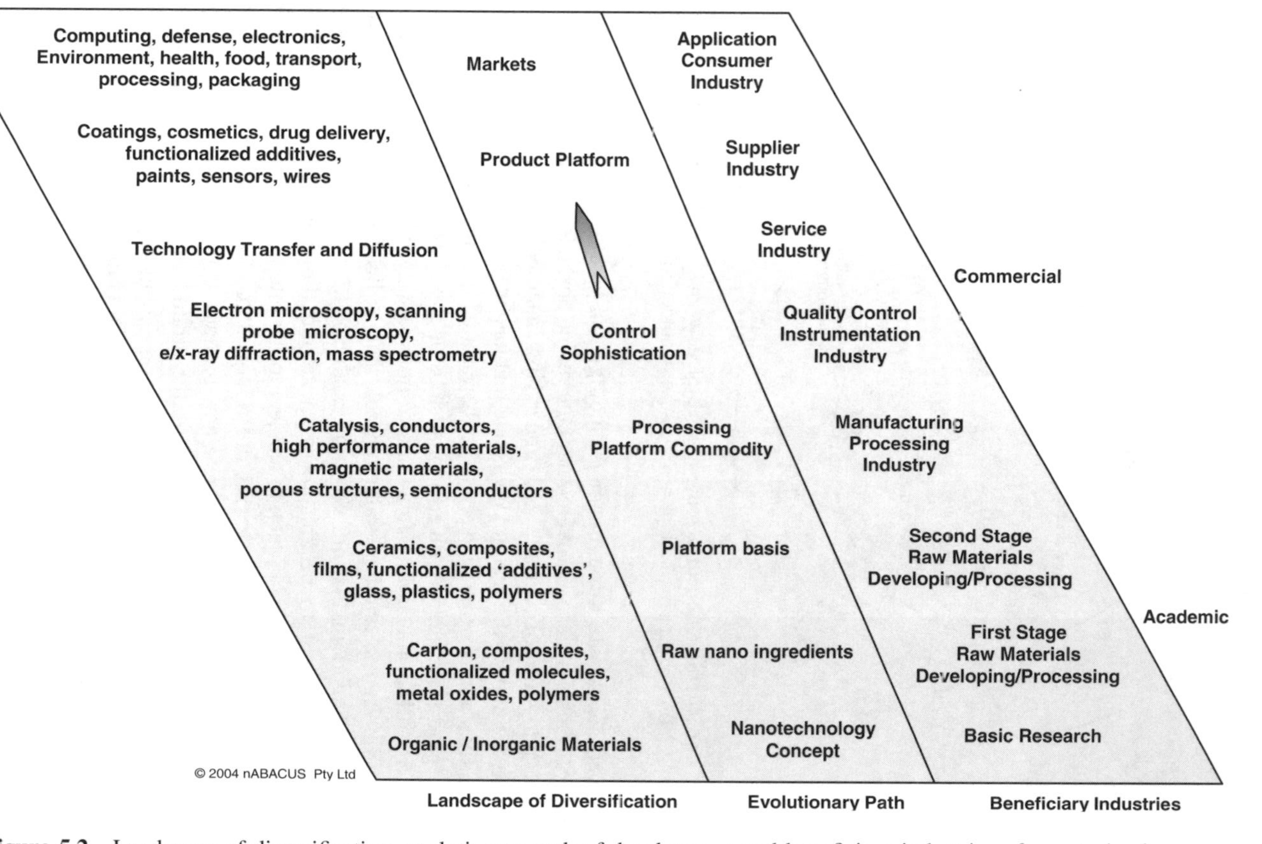

Figure 5.2 Landscape of diversification, evolutionary path of development and beneficiary industries of nanotechnology

nanotechnology concept to an application in generic markets, nanotechnology is feeding growth in many existing nanotechnology-specific industries as well as new ones. While all these industries contribute to the spread of nanotechnology and to economic growth, it can be expected that real exponential growth may occur only after the nanotechnology-enabled nanotoolboxes are ready to be used as conveniently as a Lego® block in a child's hands.

Indicators revealing the maturity of a certain nanotechnology concept and its potential impact in years to come may be derived from analysing patent developments and their trends. Nanotubes are again an ideal example for study due to their rapid cycle from discovery to the first application prototype. Figure 5.3 shows the

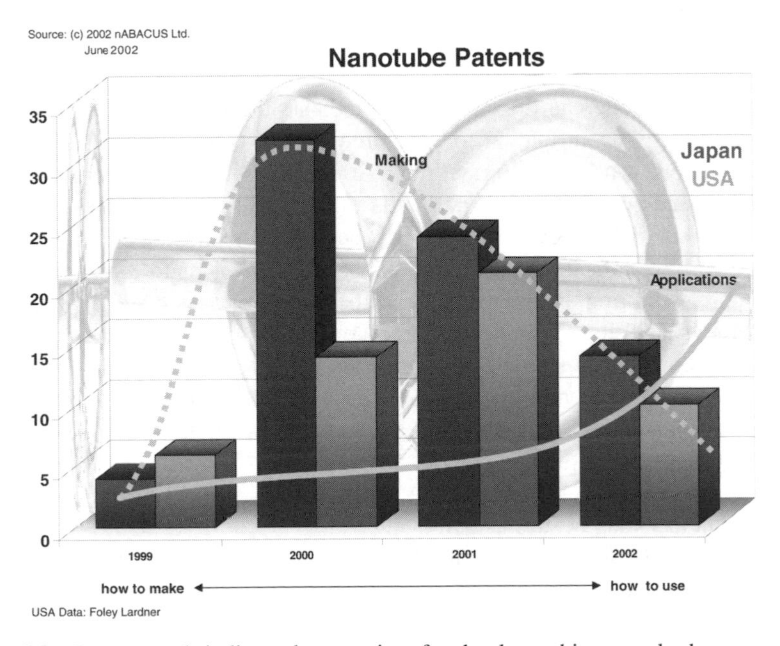

Figure 5.3 Patent trends indicate the maturity of technology: this example shows nanotubes

evolution of nanotube patents between 1999 and the second quarter of 2002. Nanotubes as a technology platform became mature, i.e. released from the lab stage and from speculation about their immediate value, when the trend of accepted patents turned from 'how to make' to 'how to use'. This occurred in 2002, after only four years of patent history.

The larger cycle in development of a technology concept and its exploitation can be observed in the three characteristic waves of patent applications (Table 5.2). Only a few patents may be able to maintain or increase their value through all three evolutionary phases, or waves. Those original patents that take part in the third and larger wave of patent applications will be the winners in the race for the best investment. Looking at the industries involved in nanotechnology (Figure 5.2) and at the general expectation of technology maturity and their accompanying patent

Table 5.2 The three phases in an ideal cycle of a nanotechnology concept to the full exploitation of its capacity and value

First wave	Second wave	Third wave
Moving from how to make to how to use Discovery of new capability and materials Refinement of new capability and materials Replacement of old capability and materials in old things	Technology replacement	Nanotechnology-enabled nanotechnology Invention of completely new products

application waves, one can draw a chart to indicate the major steps and the extent of nanotechnology's contribution to economic growth (Figure 5.4). Basic research and development work is contributing a constant stream to economic growth, although its tangible value in real dollars is difficult to estimate, but with sufficient amounts of nanotechnology raw materials becoming available, and with product processing and manufacturing in place, standardization of nanotechnology-enabled building units will bring us to a stage where the 'nano-RJ45' is taken for granted and nanotechnology will have penetrated every step of our lives.

While the entry costs to get into nanotechnology at either the development or commercialization stage are relatively low, the investment risks are as high as they are with every other seed or early-stage investment. There is the risk that

- a due diligence process does not reveal that the capability of the underlying nanotechnology has been overestimated;
- competing technologies overseas have been underestimated or overlooked (Australia, China, Korea, Taiwan);
- existing patents in related and unrelated areas could be infringed;

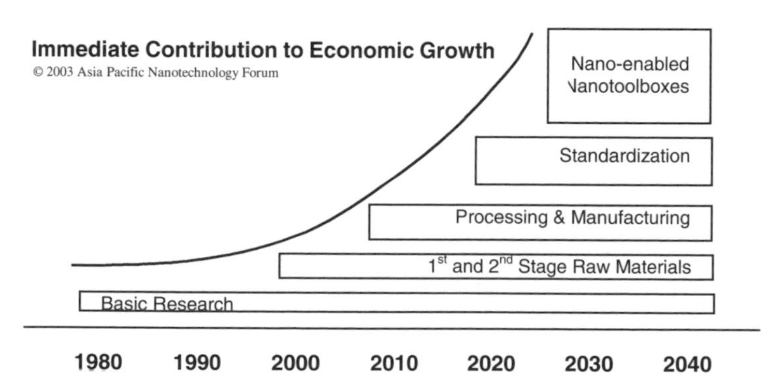

Figure 5.4 Immediate contribution to economic growth

- problems occur in scaling up production (maintaining quality, manufacturing regulations, industry and environmental standards, non-standard machinery);
- supply of raw materials shifts due to changes in regional or global demand;
- the market size and anticipated market share change during development;
- anticipated early adopters of technology have found alternative solutions;
- changes in available technologies and customer demands, and changes in industry standards and government regulations can make the product obsolete before it is even completed;
- the technology trend or adoption of technology moves rapidly in a different direction (e.g. Beta video technology lost out to VHS);
- other disruptive nanotechnology may make current technology concepts obsolete;
- the technology or product turns out to have unforeseen and undesirable side effects.

The risks when investing in nanotechnology are fairly standard and the above list is by no means exhaustive. With respect to investing in nanotechnology, there is an important observation that needs to be made. Unlike many of the expectations that lead up to the past dot-com frenzy and its rapid fall, nanotechnology has solid tangible products and product platforms to show at the end of the development and their success does not depend much on anticipated consumer behaviour. Many consumers may not even be aware that a product has been nanotechnology enhanced or that its functionality is based entirely on nanotechnology.

The success of nanotechnology products hinges, like many other new platform technologies, on adaptation by industry and the pace at which a mechanism for quality control and standardization can be achieved. The success of nanotechnology and its contribution to economic growth are very much in the hands of the industry, its professional association and government policy makers. The end consumer is merely the beneficiary of very much improved products. Naturally, the end consumer will choose the better-performing construction material or the light, comfortable and more attractive shoes, but that has always been the case in the past and is not expected to change in the future.

6

Need for a New Type of Venture Capital

Po Chi Wu
California venture capitalist

The premise of this chapter is that the role of venture capital is changing because more resources are needed to support entrepreneurial ventures today than ever before, because of the complexity of the technology, decreasing product development cycle times, accelerating global competition, and an increasing need for more capital. Some of these issues are of particular significance to companies developing nanotechnology, some are a consequence of the overall evolution of industrial development. Accompanying the most recent wave of technological progress, venture capital has been seen as a driver of innovation.

Studies by Professor Joshua Lerner of the Harvard Business School have shown that, in aggregate, the investment by corporations in venture capital has resulted in a fivefold greater impact on innovation, compared to investment in in-house R&D. This impact was measured primarily in terms of patent filings and other intellectual property.

What this means is an increasing dependence by large corporations on finding and acquiring new technology and new products developed in small, entrepreneurial companies. This trend is evident in all industry sectors, as exemplified by huge success stories like Cisco Systems, in the communications business, or any of the major pharmaceutical companies, like Eli Lilly. These kinds of companies invest in outside venture capital funds and have in-house direct investment programmes, in addition to the acquisition activities. For them, investing in new technology

Nanotechnology: Global Strategies, Industry Trends and Applications Edited by J. Schulte
 ISBN: 0-470-85400-6 (HB)

companies is like outsourced R&D. This trend has profoundly changed the venture capital environment.

The economic rationale for this direction taken by major corporations is clear. Big companies, with their huge marketing budgets and infrastructure, can do things small companies can't. Their own R&D staff can stay focused on research in their core strategic areas, and look for the most direct ways the company can add the greatest value to the overall process of getting products to the customers. What start-ups do well, big companies do less well. The creative mix of innovators from multiple technical disciplines is easier to put together in a venture capital backed start-up than in a major corporation. The start-up is, in many ways, a more effective environment, more stimulating, and often more rewarding for these people. Just having outside investors, like venture capitalists, provide capital for the early development phases is like off-balance-sheet financing for the major corporations. The start-up environment also lends itself to the intense dedication that often achieves valuable results much faster than in a large corporation. At the end of the day, the question of who does what best revolves around the issue of value, how value is best created, by whom, in what time frame and with what resources.

This scenario of outsourced R&D has particular significance for nanotechnology, where innovation is clearly being driven from the academic laboratories. Universities today are very different from what they were just 10 years ago. For many reasons, budgetary as well as social, they are much more open to collaboration and eager now to collaborate with industry. Professors today typically have consulting contracts and grants from large corporations, from all over the world, to support their basic research. Professors themselves are more interested in such opportunities, and often research is being directed at practical applications of science, and not just theoretical projects. They are collaborating on specific projects with colleagues from other disciplines, in an impressive and aggressive manner which is also very new. This trend is reflected in powerful initiatives such as the California Nanosystems Institute (CNSI) based in UCLA and UC Santa Barbara, initiatives which have received hundreds of millions of dollars in funding from the federal and state governments as well as private companies. Many other states have similar programmes to encourage nanotechnology research.

Venture capital is one of the essential ingredients for certain kinds of entrepreneurial activity. Not every type of business idea needs or is appropriate for venture capital, but for the right combination of a specific concept and particular investors, the outcome can be extremely positive for everyone concerned. This has been and will always be true. The fundamental interaction of the business involves highly skilled people aggressively focused on achieving a powerful outcome characterized by a substantial return on investment. As the venture capital industry, which is relatively new at about 40 years of age, continues to evolve and mature, what is changing is the nature of the people and how they can work together with their portfolio companies most effectively.

Venture capital today ranges from very early stage (seed) investors who are willing to work with an entrepreneur from the beginnings of a business concept to private equity investors who put their money into companies that already generate revenues and profits, where there is a reasonable expectation of a liquidity event, an acquisition or an initial public offering (IPO) of its stock, within a relatively short period of time, typically a year or two.

The term 'venture capital' has long been associated with the concept of risk capital, money that is used in situations where conservative bankers would fear to tread. Venture capitalists do not think of themselves as risk takers. They prefer to see themselves as risk managers. Risks and opportunities need to be identified and articulated clearly before a decision to invest is made. Venture capital investors can be notoriously single-minded in their pursuit to manage the risk in the investments they make. Thoughtful investors explore technical and business issues until they can see the boundaries of their knowledge. They extend their capabilities by relying on others who are truly expert in their fields.

Over the past 30 years, venture capitalists have had a long spell of success investing in new technology, punctuated occasionally by downturns in the economy. There will always be business cycles in every industry. Venture capital is no exception. The internet bubble, however, has some unique characteristics which are worth noting. The most important is that, as the economy was roaring through the 1980s and 1990s, a lot of very bright young men and women were being educated in the finest technical and business schools around the world. When these people entered the job market, they brought extremely high expectations for themselves and for the world they were going to create.

Some of them, attracted to the large sums of money being invested and realized in the stock markets, became investment bankers and venture capitalists. At the time, as money poured into venture capital firms, bright young people were being hired into the industry. Often they had had relatively little operational experience, compared to the traditional profile of venture capitalists as grizzled veterans who had proven themselves successful survivors in corporate wars. It is an ironic observation that this industry, which prides itself on investing wisely in people, in people who have proven track records, eagerly snapped up untested young talent, offering amazing compensation packages. Around Silicon Valley, the population of BMWs probably doubled, if not more, resulting in more traffic congestion and overcrowded restaurants.

In the euphoria of concept IPOs, companies were going public with their stock offerings at a time when they didn't have any immediate or even near-term prospects for achieving substantial, if any, revenue or profit. Companies like Amazon.com boldly predicted they would operate at a loss of hundreds of millions of dollars for at least five years before showing profitability. All of this excess was an unholy alliance between venture capitalists, investment bankers, securities companies, institutional investors, virtually everyone in the food chain. It really was a time when no one dared proclaim, 'The emperor has no clothes.' It was in everyone's best interests to feed the hype machine because that was the most effective way to keep the valuations and expectations rising. The more the better. The faster the better. The bigger the better.

Many venture firms which used to emphasize working closely with their portfolio company managements, building fundamental value for the business, and focusing on achieving revenue and profitability with a minimum of capital, poured money into these companies, sometimes in a competitive frenzy that drove valuations unreasonably sky-high. How did an individual venture capitalist find the time now to sit on the boards of directors of a dozen or more companies? Funds would invest in 20–30 companies in one year. What happened to due diligence?

The obvious result of this kind of activity has been devastating to overinflated egos and portfolio valuations, which needed some reality adjustment anyway, and "in terms of lost wealth" measured in trillions of dollars.

At the beginning of the twenty-first century, now what? What are venture capitalists going to do? The experienced ones are going back to basics. Firms are shedding people, returning money to limited partners, focusing attention once more on sustainable growth through solid revenue and profits in their portfolio companies. Investment rates have slowed considerably, even though valuations have dropped back to reasonable levels. Are they still investing in new technologies? Absolutely.

When will venture capitalists invest more aggressively in nanotech start-ups? In what directions are major corporations moving, the ones that will develop the major market opportunities? When is the public stock market going to accept new offerings? There are many questions, few answers. Especially given the overall climate of caution, venture capitalists are investing in only a very few selected opportunities.

6.1 The Challenges of Nanotechnology

When compared to previous waves of technological innovation, the science and business of nanotechnology present bigger and deeper challenges. Technologists are only just beginning to scratch the surface of what may be possible. The real question from an investor viewpoint is: How much uncertainty are we prepared to accept and in what timeframe? What is new about this technology is that some researchers are looking forward to practical implementations that may be 20, 30 or even 50 years into the future. We can be induced to believe that these dreams can become reality because we, especially in Silicon Valley, are believers that technology is a driving force for change. We have seen how dramatically and how quickly new technologies can be realized. We have witnessed and participated in this kind of growth and change.

To understand the impact of nanotechnology, consider the following way of looking at the multidimensional aspects of this field. There is the scope of technical *discovery* – broader and more fundamental than ever before. New discoveries can be initiated from any of the science or engineering disciplines, not to mention from industrial design considerations. There is the matter of *timing* – global competitiveness drives a very real sense of urgency. There is the *range of talents* being enlisted in research and development efforts – teams of highly experienced scientists from different disciplines all over the world, often working cooperatively. There is the matter of *global impact* – every industry in the world will be affected. Finally, there is the acknowledgement that *human creativity* is the key factor in developing new products and services that contribute value to society.

The input of creativity from teams of business people, combined with the technology innovators, has unprecedented importance at the current stage of development. This creativity derives from and feeds on the tremendous energy of obsessive scientists and entrepreneurs who are eager to share their discoveries with the world. No wonder there is so much hype! Promoters of nanotechnology describe commercial possibilities that deliberately inflame the imagination and excite emotions, positive and negative, in order to achieve visibility and prominence. No wonder many investors are confused.

6.2 What Does Nanotechnology Mean to Entrepreneurs?

Here is a statistic that surprises most people. During 2001, US-based scientists published more peer-reviewed, technical articles that their peers from any other country in the world, about 300. That is consistent with most expectations. Which country is number two in publications? China, with around 140. Japan and Europe are a close third and fourth, respectively, but China's growth rate is accelerating at an impressive pace. Where were the Chinese researchers just five years ago? There are several lessons to learn from this bit of news, mostly about what is happening in China, but also some insights into how nanotechnology as an industry develops.

Since much of nanotechnology at this stage is really about the development of new materials, the underlying science is applied chemistry, in inorganic, organic and even biological systems. Most innovative are the sophisticated computer tools and the engineering ingenuity that transform dreams into reality. Many of these new materials are assembled out of existing building blocks, but are put together in novel ways, with properties that are often designed in, deliberately and consciously created to have specific, desired physical and chemical properties. Very expensive equipment, like a tunnelling electron microscope or a clean room, is needed at certain stages of this R&D, but each lab does not need all the equipment all the time, so a lot can be shared among many researchers. This dramatically changes the economics of the R&D. This is one of the secrets to the rapid growth in China. By the way, the Chinese are also beginning to apply more aggressively for international patents on their scientific discoveries.

Virtually every country that invests in advanced R&D has targeted hundreds of millions of dollars for nanotechnology research. In 2001 about $400 million was invested in the US, closer to $500 million in Japan, and about $200 million in Western Europe, with about $400 million in the rest of the world, and the total is almost double the amount in 2000. Why such an acceleration of investment activity? The US National Science Foundation predicts a trillion-dollar worldwide market for nanotechnology products in 2015, with the largest being in materials (over $300 billion), electronics (approximately $300 billion) and pharmaceutical applications (around $200 billion). Hype notwithstanding, there is a general acknowledgement that the ability to measure, control and modify matter at the nanoscale, i.e. at the atomic and molecular level, results in materials with new structural and functional properties that have the potential to revolutionize existing markets or create whole new markets. No one wants to be left behind in this global race for economic wealth creation.

The entrepreneurial activity in China may indeed be a very significant wild card, one that entrepreneurs in the US and in other countries need to watch carefully. They are paying a lot of attention to our work – are our researchers doing the same? The educational system in China is so competitive that those who successfully achieve academic positions may very well be smarter and better educated in some ways than our technologists. Also, there is a cultural bias toward working on near-term, practical applications that may require relatively little capital to develop. The conditions of the local markets in China may mean that this approach can pre-empt some of the more sophisticated technology being worked on elsewhere.

In general, technologists are eager to form companies to develop their research, but often they are not the best people to be entrepreneurs, in the sense of business builders. This is even more true in the field of nanotechnology than in other areas of science or engineering, because the new discoveries are typically only parts of a complete solution, of a real product that could be sold. Expertise across multiple scientific disciplines often needs to be included in the team. Sometimes the technologists don't even realize the limitations of their research, such as cost-effective manufacturability, integration into existing systems or what effects their technology might have on the competitive landscape. The projects that will have the best chance of success will be led by management teams that combine business talents and experience with previous commercially successful projects. Forming the initial management team is always the highest-priority task of an entrepreneur.

One of the biggest challenges for all technical entrepreneurs is often identifying and developing the best commercial opportunities, the ones that will be most achievable and lucrative. In nanotechnology this is exacerbated because of the complexity of the technologies, and because there are many ways in which technical innovation can be commercialized. Some applications are simply not appropriate for development by a new start-up business. Typically, these would be applications in very big markets controlled by an oligopoly, such as the aerospace or petroleum businesses. Improved structural materials that find application in aerospace due to their novel physical properties, e.g. strength, weight, electrical and heat conduction characteristics, need to be sold to the handful of companies that are the principal vendors, e.g. Boeing. Boeing is very actively looking for and incorporating nanotechnology-derived materials.

In this type of situation, the start-up entrepreneur may find a licensing strategy to be the best way to generate income from the technology. Obviously, the key word is 'strategy', where the licence provides optimal benefits to the entrepreneur, including issues like exclusivity, non performance according to the original co-development agreement, and non-competition.

Entrepreneurs in the nanotechnology field need to build management teams of greater breadth and depth than in other engineering or scientific fields. The main reason for this is that the breadth of potential opportunities is so much greater than before, in many diverse markets with which the entrepreneur may not be familiar. One of the most common issues that entrepreneurs need to address in their plans is a lack of sufficiently detailed understanding of potential market niches. The niche concept is very important. A start-up doesn't have the resources, human or financial, to pursue broad markets. Not only does the technology have to deliver on its promises, but the entrepreneur has to focus all their energies on a niche where their start-up company can be the dominant player. Being the number one or number two vendor creates real value for the company, even in a small niche. Arrogantly planning to take on multibillion-dollar competitors is often a recipe for disaster. Just look at all the software companies that felt they could compete with Microsoft.

Especially in today's economic environment where financing is scarce and difficult to obtain, nanotechnology entrepreneurs need to demonstrate the practical viability of their products. Fortunately, attitudes regarding academic research have changed quite dramatically over the past decade. More research is collaborative, across multiple institutions as well as across departments. Professors are able to secure grants that will support them, their postdoctoral fellows and graduate students in doing research that essentially results in breadboard prototypes which demonstrate the technical features of their technological innovations. Universities are becoming more aggressive and liberal about licensing to new companies the intellectual property developed on campus. The good news for everyone is that these changes are favourable to the formation of new companies. Virtually every nanotechnology start-up now has its roots in academic research. The in-house corporate research at major corporations is another matter entirely. Companies such as HP, IBM and Lucent are spending hundreds of millions of dollars, if not billions, on research at least related to nanotechnology. At the same time, they are close to academic researchers and often fund their work.

Overall, innovation in nanotechnology creates very exciting opportunities for business entrepreneurs, but they need to consider many more issues than in other areas of technology.

6.3 What Does Nanotechnology Mean to Venture Capitalists?

Venture capitalists today are more cautious than ever before, not only because of the internet bubble, but also because nanotechnology is still in its infancy. Healthy scepticism is a trademark of the profession, one that is valued more these days than in the boom times. Most in the venture capital community are still focused on digesting their earlier investments from previous years. By definition, the amount of new investment per year is only a small fraction of the total amount of funds under management. Very few new funds are being formed that focus on nanotechnology.

The top-tier venture capital firms are always alert to the possibility of new breakthrough technologies and will fund a very small number of start-ups. Start-ups need to rely on the few smaller, specialist firms that have the resources to support a nanotechnology company.

Significantly, the history of investments in advanced materials has not been encouraging. Licensing as a primary business model is, in general, not appealing because companies that only license out their technology will generate very attractive profits but not large revenue streams. Such enterprises don't grow to become big companies which would be valued highly by the public stock markets, so an initial public offering of stock is unlikely to be the exit strategy of choice. More often, the technology and the company are acquired by a major corporation that may be a strategic partner initially, or even a competitor. The path to tangible value creation is long and arduous.

This situation with nanotechnology start-ups actually poses a bit of a quandary for venture capitalists. They can clearly perceive the substantial market opportunities, but they cannot clearly see the path to commercial success. The increasing multitude of factors faced by entrepreneurs are shared by a venture capitalist who is willing to invest. Venture capitalists are most comfortable when they understand the business model, one that is based on previous successful experience, but which also includes new features to deal with the current risks. Today, venture capitalists tend to be specialists in certain areas of technology: software, semiconductors, wireless technologies, biotechnology, medical devices, etc. Venture capitalists rely on real-world experience as the most valuable asset of a management team, so they must apply that same rule to their partners, if not themselves.

Who are the experts in nanotechnology? Partners with a strong background in the chemicals or materials industries are relatively rare in the venture capital world. The situation is reminiscent of the early days of biotechnology. In the days of Amgen and Genentech, there were only biologists, biochemists and pharmacists, as well as chemists and chemical engineers. What was a biotechnologist? Today the term is broad enough to include all the earlier specialists, plus people who look at the business from a different, unique point of views.

Before venture capitalists will invest large amounts of money in nanotechnology start-ups, they will have to develop not only technical expertise, but more importantly, a new and different point of view about their investment strategy. At this point in the development cycle of the technology, the smart investors are looking at powerful technology platforms with broad patent coverage of multiple market opportunities. Some nanotechnology start-ups have more than one hundred patent applications, with some already issued. Again like the biotech industry, establishing strong intellectual property positions is going to be extremely important for start-ups, for defensive as well as aggressive strategies.

Nanotechnology start-ups are more difficult to evaluate, in terms of technical performance and in terms of commercial potential. This means that venture firms may have to rely, as they currently do, on outside consultants to help them understand competitive positions. Eventually some level of this experience has be brought into the venture firm itself, because the real work of a venture capitalist happens after the investment is made.

Adding value to the nanotechnology start-up is also a greater challenge than working with companies in traditional industries. The main reason is the greater diversity – in the people, in the market opportunities, in the complexity of paths to commercial success. Venture capitalists considering investments in the nanotechnology space need to be prepared to invest a lot more than money – their own time, their creativity and their contact networks are more urgently needed. Venture capitalists who specialize in start-ups are quite different from private equity investors who focus on companies that already have market presence, and are often already profitable or nearly profitable. The operating strategies that help restructure financial alternatives in a near-profitable company are completely different from those that a start-up needs to develop strategic alliances for co-marketing, for example.

Particularly in the area of nanotechnology, I can see a return to the small, collegial syndicates and working styles that characterized the early days of Silicon Valley venture capital. Start-ups, like small children, need consistent and persistent support, whether they really understand they need it or not. That is the venture capital perspective. Ironically, one of the worst things that can happen to a company is to receive too much funding at one time. The result is often a kind of indigestion, and certainly, quite a bit of wastefulness, human nature and egos being what they are. When facing great uncertainty and unforeseeable challenges, venture capitalists like to have co-investors who share their values and are willing to work together with the company to develop winning strategies. This is not only for moral support.

Venture capitalists who invest in nanotechnology start-ups will need to focus on all the challenges faced by the entrepreneurs, as mentioned above. The biggest risk for an investor is always: What will happen when the company needs to raise more money? Who will invest then? What accomplishments will result in the creation of enough value that new investors will want to participate, and at a higher valuation than previous financings? When will the public stock markets accept new offerings? Because of the newness of these companies and their business models, there isn't consensus yet on what kinds of milestones are the most appropriate for judging the progress of a nanotechnology start-up. The standard ones that emphasize business goals, e.g. marketing or distribution deals, strategic alliances, acquisition of revenue-generating customers, are best but they might not happen during the first few years.

This lengthy period of early funding is a major reason why, from an investor perspective, it is critically important that a nanotechnology start-up is very clear about its intellectual property and how it plans to capitalize these assets. During this time, what are the milestones that add value to the start-up? What be the venture capitalist's role in this phase of activity?

Venture capitalists are often criticized for their focus on return on investment and exit strategies, but that is definitely the most important aspect of their job. They are paid to make money for their investors. If a venture capitalist cannot believe that the return on an investment is sufficiently attractive, they should not make that investment. So exit strategies are very important topics to discuss before the investment is made. Seasoned entrepreneurs may have the experience to propose various business strategies that will support multiple exit opportunities. Only the naive believe that every company can go public at some point in its growth. The big questions are: What builds the most value for the company? And how do we create that value?

Venture capitalists, by virtue of their experience and contacts, can usually be very helpful to their companies when it comes to exit strategies. Nanotechnology companies represent a new category, so no one knows today how these businesses will be valued in the public stock markets. Acquisition as an exit strategy is easier to understand, but is often less attractive, at least so far. Venture capitalists should be and usually are very sensitive to what builds value because that is at the core of their profession. If there is one thing they need to do well, beyond particular areas of expertise, it is to recognize, quantify and enhance value.

6.4 How Value is Created through Intellectual Capital

Venture capital is a mechanism to capitalize on assets arising from human creativity, ingenuity and drive. Assets come in a lot of different forms. So-called hard assets are tangible and have a physical reality that makes them easy to value by conventional standards. Real estate is valued by comparison with other pieces of property that have similar characteristics: size, location, usage, etc. Soft assets typically include people, their ideas and intellectual property, most often reduced to practice in patents and patent applications. The softness comes from the difficulty of measuring the value. Based on the contents of filing documents alone, how does one value one patent, as compared to another? The answer is that you can't.

The commercial value of a patent depends heavily on its context. Possible conflicts arising from other patent filings, current or future, are one obvious factor. How easy will it be for other inventors to work around the patents, to find other ways to achieve similar results without infringing? Another factor is the size of the true market potential for new products based on that patent. Are the market opportunities large enough to warrant further investment? The cost structures for the product development, manufacturing, marketing and distribution, i.e. the business issues, are critically important issues that will determine whether a company will find enough value in a patent. Is there a way to make money, i.e. profits, from this patent?

These issues, which define a form of value creation, comprise and result in a new class of asset, intellectual capital, which requires investment by companies and outside investors. Some aspects of this concept are not new, of course. Smart business people are successful not just because they have more knowledge, but because they are more clever in finding ways to use their knowledge in the context of the marketplace. This is another obvious truism. However, as applied to knowledge-based industries, like nanotechnology, it may be useful to work with this kind of value creation in a more conscious, deliberate and methodical way. Specifically, venture capitalists need to place more attention on this process and find ways to add value.

The formation and development of intellectual capital is really a precondition for a successful venture capital investment. How else can an investor get a sense of potential return on investment? What is the meaning and significance of initial valuation, at the seed stage, when there is little more than some ideas, patent filings and a handful of talented people? How is a business plan compelling? The compelling nature is not simply a matter of persuasive presentation skills, although those are certainly very important. The basic story of how the business will make money must be credible and powerful enough to temporarily suspend disbelief. It must be convincing and appealing enough that the investor is willing to accept the challenges presented by the risks. In short, the inspirational quality must be strong.

Only after the story is perceived as credible will the next level of questions be addressed, such as questions on whether the management team can execute the plan. Those questions may be even more important, but the first hurdle is getting attention from the venture capitalist so they will spend some time and effort thinking about the project in a positive way.

For the venture capitalist, as well as the entrepreneur, this is a kind of value element, a critical piece of infrastructure that is necessary. Many participants who come from outside the organization are needed for this infrastructure – government, academia, business people, lawyers and investors – because the value created is uniquely a collaborative reflection of human desires and needs. The venture capitalists who will be successful investing in nanotechnology will have to learn and practise this aspect of the art. How is the building of this asset, which is just another name for 'value element', actually accomplished?

Many developing countries, e.g. in Asia, have studied the models of successful venture capitalism, primarily Silicon Valley, and have tried to implement the same features. The intellectual capital aspect may have been overlooked, however, as this tends to be taken for granted in California. Conventional wisdom holds that Americans are particularly strong in marketing and sales of all kinds of products and services. Just look at the power and global reach of advertising, films and MTV, not to mention other kinds of entertainment. Why have these been so influential? The messages conveyed are strong, clear and simple. They address what people want to hear, see and feel. The appeal is unabashedly emotional. Whether we approve of materialistic capitalism or not, the appeal is extremely strong because the messages touch people at a very basic level.

These marketing skills are only beginning to be used in technology businesses, as the products become more and more accessible to average consumers. For example, look at modern automobiles. The computing power of all the microprocessors in today's cars probably exceeds that of an average desktop computer, but do car-makers highlight this fact? No. Those technological innovations are not the reason people buy new cars. Consider the on-board direction-finding technologies, based on the Global Positioning System (GPS). The fundamental intellectual property was originally developed for military purposes and was made available to the public only recently, perhaps in the past 10 years. There are now many providers of devices and software that incorporate the mapping capability.

What is the value of the original patents? When the technology was developed, did anyone guess that someday it could be used in personal transportation? Even if that possibility had been considered, when and how would that ever happen on a commercial scale? How would anyone have known what the future value could be? A GPS unit installed in a car today costs $1500–2000, a significant percentage of the price of a new car. At the moment, the price is too high for broad availability. Some luxury automobiles, costing $40 000 or more, are beginning to offer it as standard equipment.

In a start-up focused on innovation, like nanotechnology, management has a huge responsibility to maximize the value being created for the shareholders. Much lip service is paid to the same statement, applied to companies of all sizes, including major Fortune 100 corporations. Applied specifically to intellectual property, this concept means investing a lot of effort in gathering the appropriate resources to really brainstorm thoroughly as many of the future opportunities as possible. Only after that activity is completed should a clear, laser-like focus ensue, directed at the optimal targets. For those applications that do not fit the company's unique value proposition, a term I use to describe its core competencies, management must find a way to realize some value through external relationships, such as strategic alliances.

Venture capital investors, especially those who sit on the boards of directors of these start-up companies, share this responsibility. This is an essential requirement of the art of venture capital, as opposed to the hard business skills such as financial management, manufacturing or even marketing. Where and how are such skills learned? Experience, experience, experience – and a lot of intense creative effort.

If intellectual capital is truly an asset, how is it valued? Current accounting systems do not know how to measure this kind of asset adequately and appropriately. The nearest thing we have is the public stock market, which is notoriously fickle and may lack sufficient understanding of the asset in the first place, since most investors tend to see only relatively superficial aspects of the companies and the most tangible consequences of whatever technology the company might have.

In the US venture capital market, this issue of the value of human ingenuity is the most sensitive one in negotiations between investors and founders of technology start-ups, as well as with academic research institutions that may own the underlying intellectual property. Intellectual property is a technical legal term with a very specific meaning, i.e. a legal description of a certain invention or discovery. Intellectual property is not the same as intellectual capital. The description of intellectual property is objective and scientific. In that description there is no indication of value. Other scientists may recognize the discoveries as Nobel laureate material, but that has no bearing on whether a product or service based on those discoveries will achieve any market acceptance.

Market value is defined by the market, by other humans who will buy and use the goods and services offered by the company, based on their technology. Customers, in general, do not buy technology, notwithstanding a relatively small segment, known as early adopters in marketing jargon, who love to rush out and purchase the latest gizmo. They love newness and inventiveness more than usefulness or usability. For them, a large part of the pleasure of ownership is being one of the first, the privileged few among their neighbours, acquaintances or work colleagues.

As an aside, many Silicon Valley high-tech firms have failed because their early marketing and product development decisions were based on sampling only the local market, i.e. other early adopters. This is especially fatal for consumer products, for which the critical test markets tend to be Peoria, Illinois, a metaphor for a medium-sized city in the Midwestern United States. This is how market research is done by consummate consumer marketing giants, such as Procter & Gamble. They spend millions of dollars and many months test-marketing new product ideas in carefully delineated demographic segments to understand just how consumers will react to their new offerings.

With regard to nanotech products and services, there will be higher barriers of resistance because some people have deep, unarticulated fears of all the unknown aspects of the technology. The challenge is indeed much greater, because the properties are unimaginable, and the critical elements are so small that they cannot be seen except with very powerful and sophisticated tools. Fear comes more naturally when the object of the fear is mysterious, invisible and apparently counter-intuitive. At the nanoscale, physical and chemical reactions follow principles of quantum mechanics, sophisticated rules of physics and chemistry that are very different from those of everyday life.

The goals of scientific discovery have always been to explore the boundaries of the unknown, but never before has humankind been able to create new materials at the atomic and molecular level, with properties previously unimaginable from examples of nature. We are pushing the envelope of human creativity and harnessing its awesome power to create tools that will profoundly change our world and our relationship to it. Ethical and moral issues abound in this kind of environment, as well as more mundane business issues. Global issues are evident, as teams of researchers from all over the world race to seize the initiative in establishing new beachheads of intellectual property.

6.5 Perceptions of Value

Consider the evolution of economic business models, from the Agricultural Age, to the Industrial Age, to the Information Age, and the definitions of 'working assets' in each case. What were the essential elements of these businesses? What were key measures of productivity, i.e. value?

In the Agricultural Age, the major asset for a business was raw land, which humans developed into farms, raising livestock and vegetables. Apart from developing tools, humans could only contribute their physical effort and hope that Nature would benevolently provide (manufacture) sustenance in the form of healthy harvests which could be sold. The land asset was tangible (fixed) and could appreciate, in response to increasing demand. Simplistically, one key measure of productivity value in this nature-based economic structure is return on assets (land). The value of one farmer's property could be compared to another's by the amount of revenue it could generate per acre or per square mile. A farmer buying a piece of farm property from another could reasonably expect to do at least as well as the seller, if used for raising similar crops.

In the Industrial Age, humans introduced a new class of major assets for a business, i.e. manufacturing facilities and equipment, created by investment dollars as well as technical innovation. Now, with the land and other natural resources, humans could control the manufacture of goods and services that could be sold. These man-made assets, however, depreciated with time, and had to be renewed with additional investment. Using the same simplistic train of thought, the key measure of productivity value in this manufacturing-based economic structure is return on equity (facilities + land). One measure of the value of one carmaker relative to another, for example, would be to compare the number of cars produced per square foot of manufacturing capacity.

In the Information Age, we have created yet another new class of primary asset, scientific and technical knowledge; this is intangible, constantly changing and self-perpetuating, in that knowledge naturally tends to create more knowledge, with further investment of effort. In this case, human creativity drives and integrates (human) resources and capital to create intellectual capital, which results in goods and services that can be sold. The key measure of productivity value in this economic structure, then, is return on investment of effort and creativity.

Knowledge, intelligence, creativity and effort are not new, of course. Humans have used these abilities to differentiate themselves from others since the beginning of time, in all economic structures and endeavours. What is new now is the reduced importance of the older classes of assets in determining value. Does anyone know or care how many square feet of office space Microsoft owns? What bearing would that statistic have on the value of Microsoft stock? The investment in R&D, though, and the quality of its engineers, are extremely important aspects of value for the company. What is an appropriate way to measure the value of its combined corporate intellectual capital? This is a major challenge to existing economic models and accounting principles, which are based largely on manufacturing and even agricultural business structures.

In this context, intellectual capital is a fundamental and new concept of a major class of asset in the information economy. This asset is intangible (mostly), but it can be embodied in various descriptions, e.g. patent documents or physical inventions. Unlike other physical assets, it contains the seeds and primal energy to renew itself, through extensions of creativity, through human curiosity and exploration. New discoveries and new inventions stimulate human imagination and motivate creativity. Human ambition and competitiveness quite naturally find ways to build business opportunities on those thoughts.

Over time, the value of natural resource assets, e.g. land for agriculture, typically declines with depletion, but the value can increase as demand for new applications of that land increases. Human-created hard assets, e.g. buildings and equipment, begin to depreciate once they are fabricated. Human-created soft assets are fundamentally different. Their value follows a cycle of human interest and motivation, from low to high, and back down again, and the cycle can repeat itself, just as fashion trends tend to repeat every few years.

This is a new concept of productive asset. Intellectual capital is a living, dynamic entity, not a fixed asset at all. The value is constantly changing, as it reflects changes in human understanding, acceptance and use. Intellectual capital includes objective knowledge in science and technology, plus elements of perceived value by and for humans, i.e. marketing savvy which recognizes the benefits to human needs and human society. Usefulness as a measure of value is defined as the extent to which knowledge or intellectual capital is perceived to address human needs or concerns.

Purely academic research is perceived to have little direct and short-term value to society because it is understood by only a relatively small number of highly trained individuals. Its target market is very narrowly focused, so the general public has no clear perceptions of the content or value. Unfortunately, the public at large has a tendency to discount the value of what they do not understand.

Market value is, by definition, social value, value to people in their everyday lives. Market value is created by a feedback process where the creators of knowledge must interact with the public directly and indirectly to communicate and share their understandings of (future) value and of what is being created.

Branding of consumer goods is even an example of intellectual capital. The power of this approach is validated in the hundreds of millions of dollars a major consumer products company, such as Coca-Cola, invests every year to influence and shape public opinion and choices. There are people who literally cannot live without their daily dose of Coke. This phenomenon is more than a physical addiction. These are lifestyle choices people make to define an environment that is positive for them. What does the brand mean to these people? Convenience, familiarity, social identity, personal definition? Do any of those characteristics have anything to do with the contents of the drink? Are they not more the result of persuasive stories and images presented through incessant and ubiquitous advertising? The story sells the product. Why else present images of vibrant, young, beautiful men and women in the advertisements? Because the story promises that you, too, can be like them, if you only join by purchasing the product.

Venture capital firms in Silicon Valley have generally been less successful with investments in consumer products companies than in start-ups focused on industrial customers. Why is this? For one thing, many of the older venture capitalists came from a technical background themselves, so they could appreciate the value of innovations. The investment game has tended to focus on speed, on a company being first to market, dominating a specific niche by solving a particular issue. Because these are solutions, the customer base is more clearly defined, more predictable, and more stable. Purchase decisions are, or at least used to be, the result of rational, straightforward logic based on economic and technical arguments. Consumer products are, by definition, designed and marketed to appeal to the emotions, at least as much if not more than to meet logical thought processes. In most electronics markets, regulatory hurdles are few and relatively easily met, compared to the challenges presented by government agencies like the US Food and Drug Administration

(FDA) for new pharmaceuticals. In many ways, these target markets are simpler and more approachable, with a lower capital requirement than for consumer products.

The same observations about the role of venture capital in nanotechnology would appear to be true. Using this new framework of intellectual capital, it seems that it is, or at least was, simpler to create value out of technical innovation in traditional electronics or information technology because the markets are narrower and more clearly defined. Nanotechnology presents some paradoxes. Precisely because these innovations are tremendously powerful in fundamental ways, they have potential application in many diverse industries. That characteristic makes new opportunities more intriguing, exciting and appealing to the imagination and emotions.

Those emotional responses are danger signs for a venture capitalist. On the one hand, the venture capitalist must react with some excitement before any further discussion can take place. On the other hand, these are signals that a lot of groundwork in due diligence must be done to validate and quantify the opportunities before investments are made. The good news is that there is a lot of opportunity. The bad news is that the potential for a lot of opportunity means that more time and effort will be required before the potential can be properly identified, much less realized.

The typical first response by a traditional venture capitalist is cautious waiting. 'The project is really interesting. Let's see what develops over the next few months.' How often is that phrase uttered? To the entrepreneur, this kind of encouragement is really doublespeak. It's a rejection that is not a rejection. It does represent some validation for the value of the project, but obviously not enough to be accompanied by a cheque, preferably in seven figures.

What can be done to make this process of funding innovation more effective? At the heart of such a goal is always education – for both entrepreneurs and venture capitalists. Entrepreneurs need to be more aware of what is fundamentally required to build a successful business, especially the intellectual capital and management aspects. The most common complaint from the venture capital community is that most entrepreneurs are simply not adequately prepared. From the entrepreneurs, interestingly enough, the major complaint is the same! To them, there are too few venture capitalists with the technical knowledge, vision or creativity to appreciate their projects. Even fewer have the time or make the commitment to their portfolio companies, to invest themselves, their insights and knowledge, along with their money.

The entrepreneur–venture capitalist model of building a business is clearly symbiotic, not just synergistic. Each side needs the other to achieve shared success. Both success and failure are shared. The arguments about who loses or gains more than the other are endless and not relevant here. For this kind of partnership to work in the long term, which is the reality of venture investing, compatibility among the partners is essential, meaning some kind of parity in capabilities, values and commitment. As the venture capital industry has matured, there are more candidates in both camps who know how to make the process work well. But this is all relatively recent.

Both sides have begun to recognize what they have to learn, about each other and about themselves, as well as about the technology and market opportunities. The successful ones are also learning effective communication skills. You can't work with someone if you don't communicate easily, fluently and often.

Venture capital firms are changing their operating structures to include more senior people with extensive backgrounds actually running companies at various stages of growth. Venture capitalists are becoming more aware of their responsibility to their portfolio companies. This is one of the more positive consequences of the internet bubble bursting. Being a venture capitalist is a very humbling experience in some ways. You have to live with the knowledge that the majority of your investment decisions will turn out to be poor, i.e. will not achieve much return on investment, so you are wrong most of the time. In spite of this, a venture fund can have an outstanding return on the portfolio of investments because the return on the few that are successful is so high. At the time of investment, a venture capitalist has to believe that each company has an excellent chance to succeed. Otherwise, why invest at all? There also has to be confidence that the portfolio will be successful. The problem is that there is no way of knowing in advance who the survivors are going to be.

Because academic researchers are now also more aware of and interested in commercial opportunities that can arise from their work, they are now more open to working collaboratively with venture capitalists. Traditionally, there has been quite a wide culture gap, practically a chasm, but that has changed in recent years. The negative stereotypes on both sides are no longer considered as industry norms. This is a good thing. There are still some greedy, grasping, harsh venture capitalists, but there are also more who believe in win-win propositions and treat others with respect. Although there are still some ivory tower academics, as there should be, there are also many more who have a practical bent and who cheerfully acknowledge their lack of business skills.

There is another important aspect of this alliance. How are we, as a society, prepared to deal with some of the paradoxes and challenges that nanotechnology presents to morality and ethics? Is this an issue that venture capitalists should be concerned with? How is this different from the financing of weapons research, be it nuclear, biological or conventional? Traditionally, venture firms prefer not to invest in projects where the only customer is a government agency, domestic or foreign. This has been for purely practical reasons, not on moral or ethical grounds. How much of a return on investment is possible? That has been the primary criterion.

For nanotechnology research, because of the broad scope of possible applications, this question definitely comes up again and again. In the current investment climate, where start-up companies have difficulty getting funding, some government contracts look very attractive. To the credit of the industry, various organizations, like the Foresight Institute, which promotes awareness and communication about all aspects of the nanotechnology industry, are raising these important questions.

Personally, I see some parallels with the controversy over research into cloning of human cells. Both technologies have the potential for huge impact, positive and negative, in many ways. The fundamental common characteristic that is most disturbing for lay people could be the aspect of playing God, of manipulating natural elements in ways that Nature never originally intended, with results that are completely unknown at this time. The impact on large segments of the population could be far-reaching and irreversible. To give one small example, it will soon be possible, using nanotechnology, to create surveillance systems that are so small and ubiquitous that we would not even be aware of their presence, even inside our bodies! Of course, these will have been marketed for security, to monitor the presence of bioterrorist activity, or for medical purposes, all of which would seem to be unmitigated good causes.

I personally dealt with a similar issue when a portfolio company developed wireless technology that would reveal the detailed movements of a cellphone user, including when that user was in the bathroom, as long as the phone was turned on. Our company was by no means the only one working in this direction. Should we, as an investor, have continued to support this company? On what basis would we refuse to do so? These are not simple issues to resolve.

Most venture capital firms are very pragmatic and generally stay out of moral dilemmas. They see their job as making money for their investors, as long as that is accomplished in completely legal ways. Most don't invest in entertainment, in films, artworks or gambling. Of course, the adult entertainment industry is not considered as a possibility, either. These are not just lines of moral judgment, although there might be some truth in that. More important are the fundamental aspects of the industry, what other kinds of companies you are dealing with as suppliers and customers, what kinds of partnerships are possible. These are conscious decisions a venture firm has to take about how it wants to make money. There are lots of different paths.

Nanotechnology cannot be compared to these industries. On the other hand, returning to the comparison with cloning research, it is also very evident that aggressive competition in technology development exists in many countries around the world. How can one country, such as the US, unilaterally decide to allow researchers in other countries to achieve a significant advantage? Knowledge really is like the magic genie who is let out of the bottle. The very nature of knowledge seeks freedom and expansion. Despite attempts at book burning and other forms of censorship in certain societies, our global economy and more than 6 billion people, not to mention the millions of databases, make it impossible for dissemination of information to be controlled for very long. An investor or government official can't bury his head in the sand and wish that the new technologies would just go away.

Personally, I suggest that venture capital firms develop more awareness and sensitivity to these issues. The assets being created through the intellectual capital process are powerful and life-transforming. As a society, we need to learn how to use them responsibly, while figuring out how to maximize their commercial value. Sometimes there will be trade-offs, and maybe that is also a good thing, once in a while. Challenges need to be articulated and discussed with their portfolio companies as well as political leaders.

For better or worse, venture capitalists need to acknowledge that they have leadership responsibilities to society at large. The power of the capital formation is one aspect. The power of the intellectual capital formation is even greater.

6.6 Summary

Life is definitely more complicated than ever before, regardless of what aspect we look at. Business is more challenging; there are more dimensions, more factors to consider, more unknowns that need to be addressed. More knowledge is required; more people of diverse backgrounds need to participate in the creation of value.

- *Question*: How is market value created out of innovation?
- *Answer*: To direct and manage this process, an effective approach is to use of selected human resource teams.

The teams will include multiple points of view from a cross section of human society, to assess common needs and expectations. Representatives from the following areas are encouraged to participate actively in the process:

- Public relations people deal with issues of visibility, credibility and desirability.
- Creative translators can translate innovation to different venues or contexts.
- Business people can only make money when people or customers acknowledge some benefit to them. Their commitment to buy determines value.
- Legal experts protect the intrinsic value of the innovation.
- Government agencies may regulate or promote certain kinds of innovation.

The successful articulation and communication of a business concept that integrates all of these points of view will be the basis of a humanistic approach to building a business with true long-term value.

Part Three

Frontiers of Nanotechnology

7

Frontier Nanotechnology for the Next Generation

Tsuneo Nakahara and Takahiro Imai
Sumitomo Electric Industries Ltd

This chapter proposes how to select unique research targets of frontier nanotechnology fields by considering the small size effect and the nano size effect. Examples are given for each size effect.

At the Asia-Pacific Nanotech Forum held in Tsukuba, Japan, in February 2002 more than ten policymakers from various industrialized countries delivered speeches about national strategy together with budgetary plans for frontier nanotechnology. In particular, speakers from newly industrialized countries in Asia strongly insisted that they would put the greatest emphasis on nanotechnology and increase their budget rapidly as much as possible. And they said they were quite sure they would catch up the US and Japan by the time of mass-produced nanotechnology products, as they did in the semiconductor and electronics industries.

All the policymakers said that they planned a large budget for research and development on frontier nanotechnology for several years starting in 2001. Almost all of their budgets were allocated to very similar projects previously proposed by Japan and the US, such as nanocarbon materials, nanoelectronics and nanobiomaterials. Consequently, there will be a fear that too many budgets for very similar projects will create a nanotechnology bubble that will eventually burst. It is strongly recommended that they adopt different approaches from each other so

Nanotechnology: Global Strategies, Industry Trends and Applications Edited by J. Schulte
 ISBN: 0-470-85400-6 (HB)

that they may be able to reach unique and complementary achievements in frontier nanotechnology.

How can we select unique and original research themes in the field of frontier nanotechnology? We need to consider the original target of the nanotechnology.

7.1 What is the Target of Nanotechnology?

Figure 7.1 illustrates a beneficial way of selecting innovative themes in frontier nanotechnology. The left-hand graph shows resistivity change of various wires as a function of temperature. The resistivity of ordinary metallic wire such as copper wire decreases linearly as temperature goes down. The resistivity of cryogenic wire such as highly purified aluminium wire exhibits a step change at a specific low temperature but never becomes zero, even at 0 K. Notice that the resistivity of superconductor wire decreases as temperature goes down and abruptly becomes zero at a certain low temperature, T_C, called the Curie temperature.

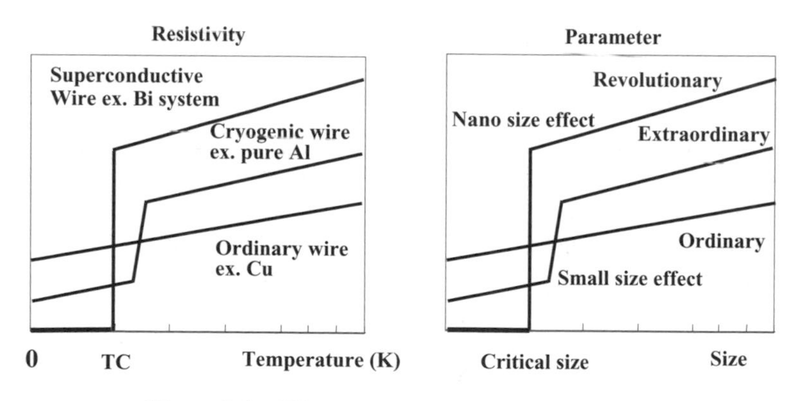

Figure 7.1 What is the target of nanotechnology?

The right-hand graph illustrates change in a certain physical parameter as a function of size, assuming that the phenomenon is similar to change in resistivity as a function of temperature. With ordinary materials the parameter may change linearly as size becomes small. With some extraordinary materials the parameter may exhibit a step change at a certain small size as in the cryogenic wire. Let us call this the small size effect. Notice that with some revolutionary materials the parameter may change surprisingly at a certain critical nanosize, as in the superconductor wire. Let us call this the nano size effect.

Figure 7.2 shows an example of the small size effect. This is the case of compressed ferrous alloy powder developed by Sumitomo Electric in 2001. The compressed alloy powder shaped like a coin shows high electromagnetic wave absorption in the microwave frequency region this is duc to resonance. By gluing

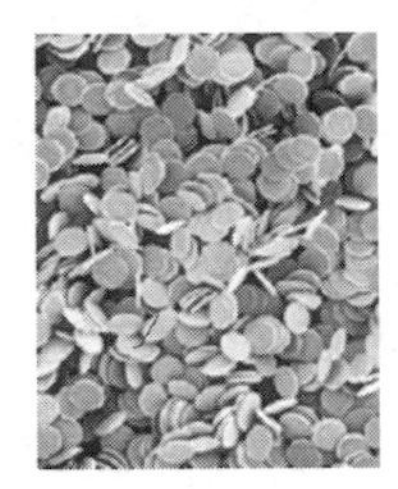

Compressed ferrous alloy powder

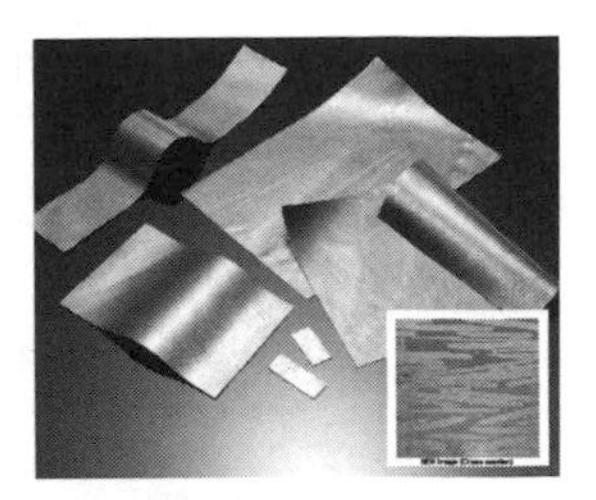

Electromagnetic waves absorber

Feature

- Particle shape is controlled at the nano-level, giving high magnetic absorption characteristics
- Adjustable particle shape and metal composition giving optimized absorption peak from 0.5 to 5 GHz

Application

- Cell phones
- Game consoles
- BS/CS converters
- VTRs digital cameras
- Personal computers

Figure 7.2 An example of the small size effect

these coins together, excellent performance has been obtained for thin and large-area electromagnetic wave absorbing sheets. These electromagnetic wave absorbing sheets are especially suitable for small and precise communication and for electronic equipment such as cellphones and personal computers.

Figure 7.3 shows carbon allotropes to explain an example of the nano size effect. Note the significant difference between a single crystal such as a diamond and a

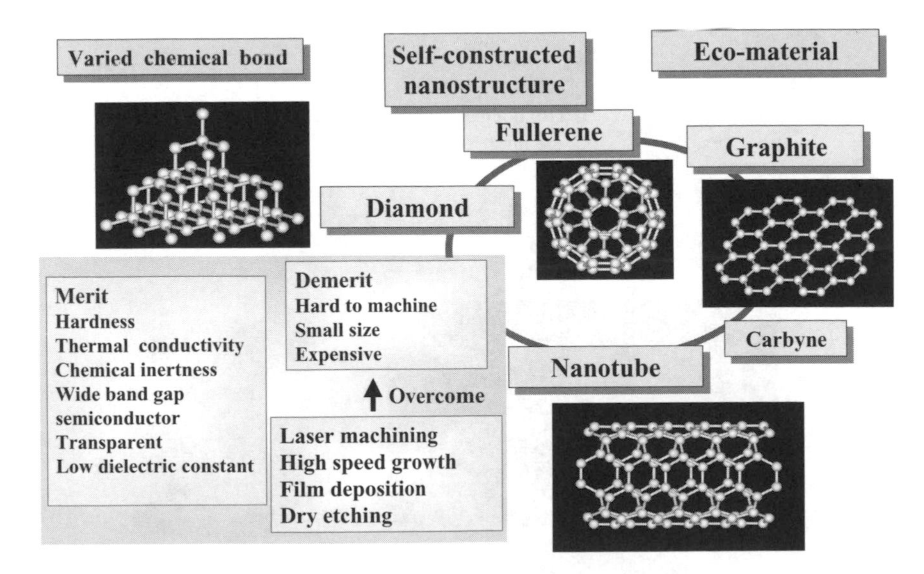

Figure 7.3 An example of the nano size effect

uniform cluster molecule such as a fullerene. Fullerenes are characterized by their atomically uniform size, autonomous formation for synthesis and quantum effective functions. Diamond is entirely different. A detailed explanation will be given below.

7.2 Diamond Nanotechnology Is a Good Illustration

Diamond has many excellent properties as a semiconductor and it can be precisely machined into a nanostructure. Diamond is not a substantially self-structured nanomaterial, unlike fullerene so. Nevertheless, there are three reasons why diamond can be considered as one of the best nanomaterials. The first is its rigid atomic structure that gives diamond an extremely high hardness, very high thermal conductivity and high acoustic velocity. The second is its properties as a semiconductor, which suggest applications for semiconductor devices, optical devices and electron emission devices. The third is the recent advanced developments in diamond fabrication and synthesis technology.

One of the most outstanding advantages of diamond as a nanomaterial is that it can be manufactured very precisely in a controlled manner. This is particularly important during precision industrial mass production such as for nanoelectronics. Sometimes the precision of products made from self-structured material like fullerenes and carbon nanotubes is very sensitive to the conditions in the manufacturing environment, just as with agricultural products.

Diamond has many distinctive properties as a semiconductor and can be extremely precisely machined on a nanometre scale compared with other materials. Even now,

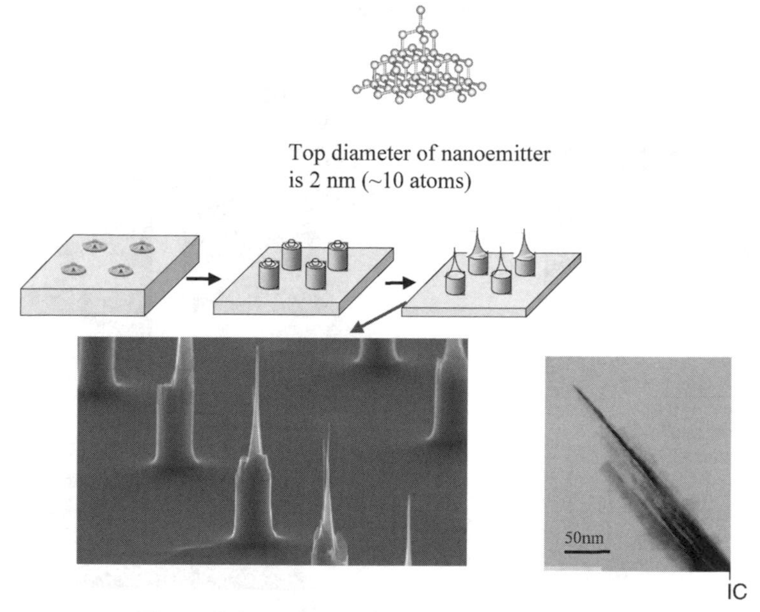

Figure 7.4 A diamond nanoemitter of size 2 nm

machined nanoscale diamond is at least comparable to self-structured nanomaterials such as carbon nanotubes. Figure 7.4 shows a photograph of nanostructured diamond. This is a steeple diamond single crystal made by reactive ion etching using patterned aluminium sacrificial masks. The aluminium mask disappears after precisely guiding the position of the steeple on the diamond surface.

The radius is 2 nm at the top of the steeple in Figure 7.4. This is nearly equal to the radius of a carbon nanotube. The size 2 nm can be considered as one of the most advanced examples of top-down nanotechnology. Figure 7.5 shows the measured electron emission from the diamond tips made by this method as compared with tips made from flat diamond. It is surprising that the electron emission from the tips was increased by almost 1 million times at the applied electric field of 1.0 V/μm as compared with that from the flat diamond surface. This can be called the nano size effect.

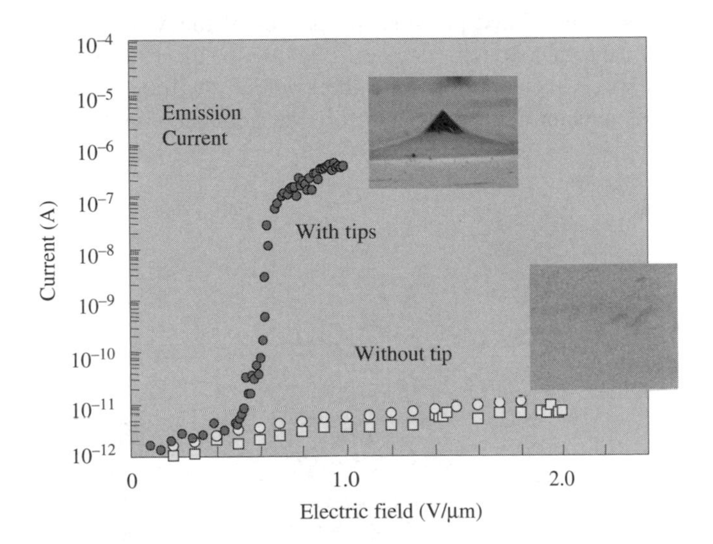

Figure 7.5 Electron emission characteristics

Many years ago the triode vacuum tube was developed and was used industrially for a long time. Then it was necessary to prepare very high temperatures of over 2000 °C in order to get electron emission from the cathode of the triode vacuum tube. Therefore the size of the triode vacuum tube was of order several centimetres. Because of its large size and its poor reliability, due to the high temperature, it was replaced by solid-state semiconductors in many places. Now, with this diamond nanoemitter, the required temperature for reasonable electron emission becomes 30 °C, which is almost room temperature. The size of the triode vacuum tube can be squeezed down to a few micrometres. Figure 7.6 shows a design example of such a micro vacuum triode. Let us call this a vacuum microelectronic device (VMD).

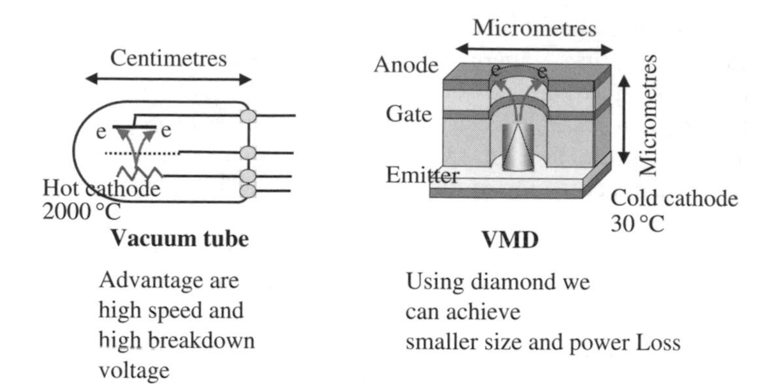

Figure 7.6 Schematic diagram of a vacuum microdevice

Figure 7.7 shows estimated potential properties of the VMD as compared with other conventional semiconductor devices. Because the electrons in the VMD travel in the vacuum space, their mobility and their velocity will be much larger than in any other semiconductor materials. Therefore the frequency limit for the VMD will

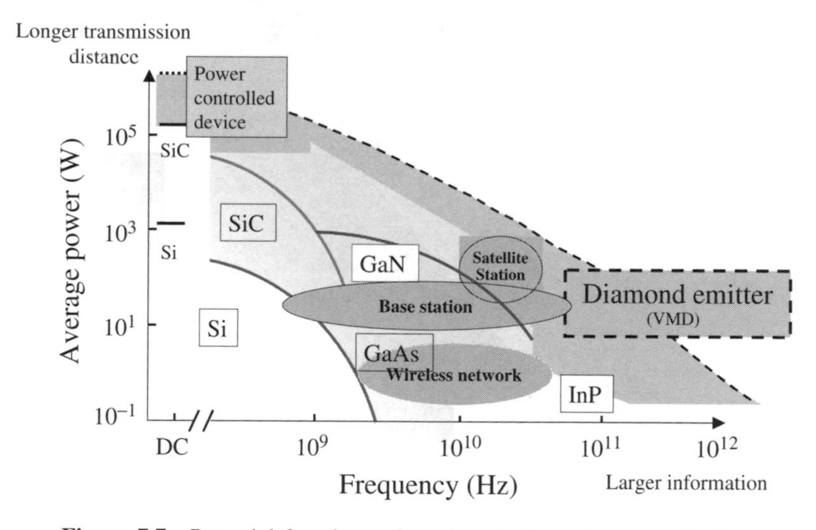

Figure 7.7 Potential functions of semiconductor and vacuum device

be much higher than for the other semiconductor devices, as shown in Figure 7.7. Also, performance of the VMD under a high-power applied environment is expected to be excellent because of the nature of diamond.

The diamond nanoemitter explained above was one result of work undertaken by Sumitomo Electric Industry, Ltd and its partners under the auspices of METI and NEDO in Japan. The project extended from this work is now a new Japanese national project that was begun in fiscal year 2003 by METI and NEDO.

7.3 Conclusion

In the field of frontier nanotechnology, there will be a tremendous number of opportunities for selecting research and development programmes from top down to bottom up. Also, there will be a considerable variety of research and development fields classified by various materials, structures and processes. Among these frontier nanotechnology projects, projects on the nano size effect discussed here will be most effective in creating the next generation of industry. A diamond nanoemitter project was explained as an example.

If huge sums of money are invested in very similar projects worldwide, there will be fear that a bubble may be created and then collapse, as happened a few years ago with information technology. It is recommended that each research institute and each industry around the world should make an independent plan for the frontier nanotechnology with their own unique programmes and perform research and development work aimed at original achievements in order to avoid duplication. When, in the near future, these achievements are integrated worldwide, the next-generation industries will be created more efficiently.

8

Next-Generation Applications for Polymeric Nanofibres

Teik-Cheng Lim and Seeram Ramakrishna
Nanoscience and Nanotechnology Initiative, National University of Singapore

8.1 Background

Polymeric fibres that possess a high degree of molecular orientation along the fibre longitudinal axis can be formed from a solution or melt of the polymer via various techniques that involve a number of molecular processes. These processes involve time- and temperature-dependent molecular motions, phase transitions under high stress, entanglement constraints and various intermolecular reactions. As a result, the final state of molecular order in a fibre depends on the process variables. These variables are stress, strain, time-dependent temperature and the length distribution of the molecules. These polymer microfibres (of diameter in the range ~1–100 μm) can be obtained using spinning or drawing techniques.

Investigation into the structure of polymeric microfibres has brought about the ability to manipulate structural formation, that has resulted in fibres of high tensile modulus and tensile strength. These high-strength high-modulus fibres are used in producing ropes, satellite tethers and high-performance sails and in polymer composites for applications such as aircraft, boats, automobiles, sporting goods and biomedical implants. In addition, a number of remarkable fibre properties include UV resistance, electrical conductivity and biodegradability. This ability to engineer properties of microfibres to meet specific requirements has resulted in

Nanotechnology: Global Strategies, Industry Trends and Applications Edited by J. Schulte
 ISBN: 0-470-85400-6 (HB)

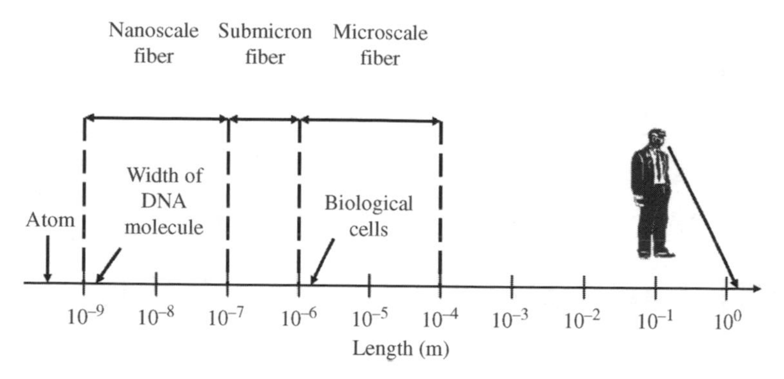

Figure 8.1 Relative diameter of micro- and nanofibres to other objects (in logarithmic scale)

demands from the industry for production of nanodiameter fibres with desired features. Structured polymeric fibres with diameters in the submicron range are of considerable interest in a variety of applications. Figure 8.1 gives a systematic view of micro-, submicron- and nanoscale fibre diameters in comparison to a typical adult height. The importance of decreasing fibre diameter can be gauged from the percentage utilization of microfibre in total fibre consumption, as shown in Figure 8.2.

The reduction of the fibre diameter to the nanometre scale (about 10^{-3} to 10^{-1} μm) results in a number of superior properties such as increase in surface-to-

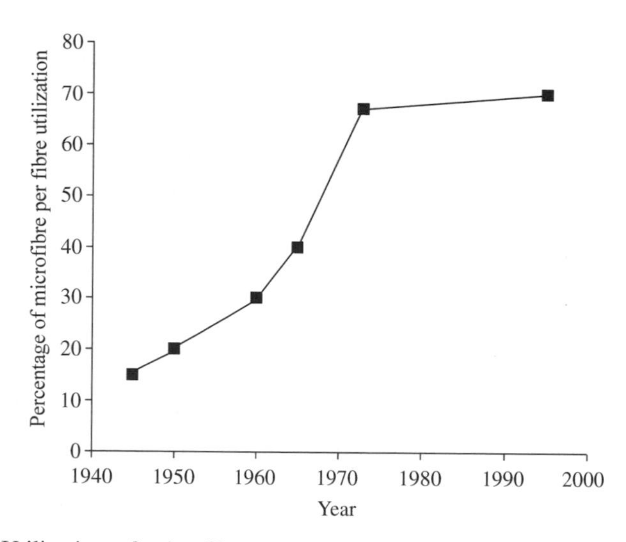

Figure 8.2 Utilization of microfibre as a percentage of total fibre consumption in the United States. Data condensed from *A Short History of Microfiber*, The Andromedan Design Company, 1998, www.googalies.com/microfsa.html

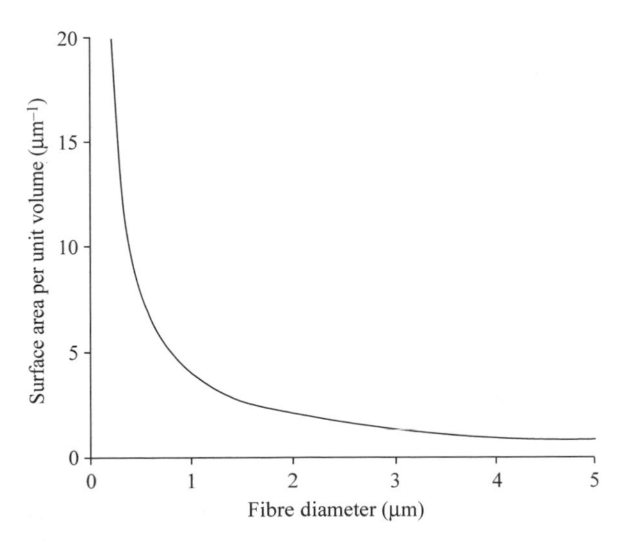

Figure 8.3 There is a sharp increase of surface area per unit volume with decreasing fibre diameter

volume ratio (Figure 8.3), decrease in pore size, a drop in structural defects, and enhanced physical behaviour. Consequently, nanofibres are excellent candidates for application in tissue engineering, high-performance filtration, chemical-biological protective clothing and polymer composite reinforcement. With the relative application of microfibre approaching its saturation point in the 1970s and 1980s,

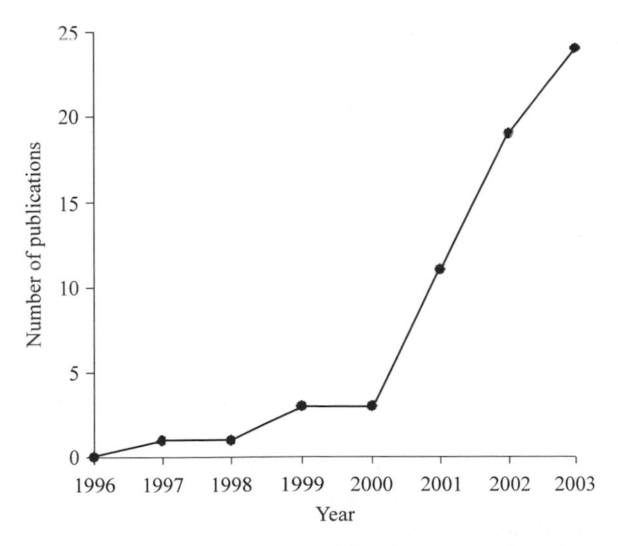

Figure 8.4 Annual number of scientific publications on polymeric nanofibres. Data obtained from Institute of Scientific Information (ISI) based on these four terms: polymer, polymeric, nanofiber, nanofibre

we may expect the rise of nanofibre application in the 2000s in a similar fashion to microfibre during the period 1940 to 1960.

Polymeric nanofibres have undergone a surge in development over the past decade. This can be demonstrated by the number of publications shown in Figure 8.4. As the research results of today may lead to commercial products of tomorrow, we take an overview of present research and development of polymeric nanofibres as a glimpse of next-generation applications.

8.2 Biomedical Applications

8.2.1 Medical Prostheses

Several US patents that describe techniques for making vascular prostheses [1–6] and breast prostheses [7] refer to the use of polymer nanofibres. Recently, protein nanofibres were deposited as a thin porous film onto a prosthetic device for implantation in the central nervous system [8–11]. The film with a gradient fibrous structure was expected to efficiently reduce the stiffness mismatch, hence the stress concentration, at the tissue/device interface, thereby preventing the fracture or fatigue failure of the device after implantation.

8.2.2 Tissue Engineering Scaffolds

Naturally occurring materials such as silk, keratin, collagen, viral spike protein, tubulin and actin (proteins), cellulose and chitin (polysaccharides) and mucin (a glycoprotein) are characterized by well-organized hierarchical fibrous structures ranging from nanometre to millimetre scale. Therefore materials for application in tissue engineering should be biocompatible with, and mimic, the native tissue structure in order to fulfil its biological functions. The challenge is to design three-dimensional scaffolds of synthetic biodegradable matrices that provide temporary templates for cell seeding, invasion, proliferation and differentiation, thereby resulting in regeneration of biologically functional tissue [12]. Biodegradable scaffolds from nanoscale polymeric fibres may hold the key in adjusting the degradation rate of a specified biomaterial in vivo. This may well be inferred from the work of Migliaresi and Fambri [13] that reveals the effect of microfibre size on degradation delay and related properties.

Due to the fact that cells gather and attach around fibres with smaller diameter than themselves [14], attempts have been made to convert either synthetic polymers or naturally derived materials into small nanofibres and nanofibrous structured scaffolds to mimic the morphological characteristics of native fibres [15–21]. It has been shown recently that smooth muscle cells oriented themselves along aligned nanofibres [22, 23] thereby suggesting a mode of anisotropic tissue growth, as shown in Figures 8.5 and 8.6.

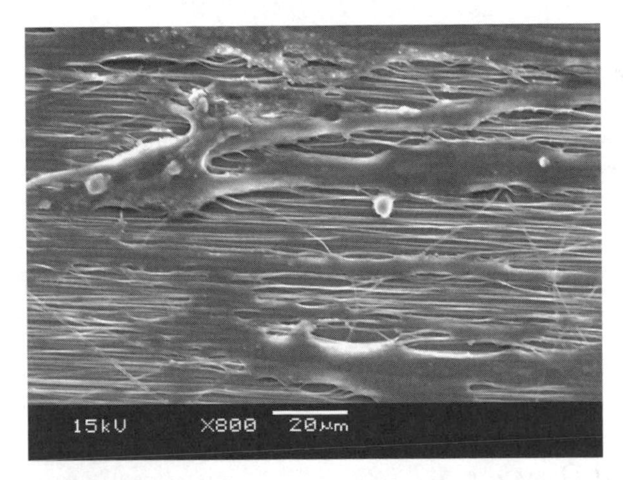

Figure 8.5 Smooth muscle cells attached and grew along the alignment of the copolymer P(LLA/CL) 75:25 nanofibre

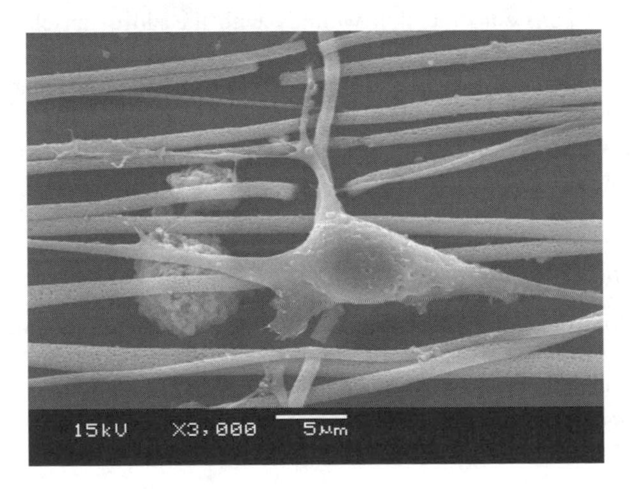

Figure 8.6 Neurites grew parallel to aligned nanofibre and branched along non-aligned nanofibre

8.2.3 Drug Delivery

Drug delivery in the most physiologically acceptable manner has always been an important concern in medicine, and nanofibres have been demonstrated to exhibit promising potential in this area. Drug delivery via polymeric nanofibres is based on the principle that the dissolution rate of a particulate drug increases with increasing surface area of both the drug and the corresponding carrier if needed. Recent progress includes electrospun nanofibres for pharmaceutical compositions [24], tetracycline hydrochloride based on fibrous delivery matrices of poly(ethylene-*co*-vinylacetate) and poly(lactic acid) [25] and biosorbable nanofibre membranes for loading an antibiotic drug, Mefixin [26]. Since the drug and the carier can be mixed together in the resulting electrospun nanofibre, the following next-genetration drug-delivery nanofibres can be expected:

- drug particles attached to the surface of the carrier, where the carrier is in the form of nanofibre;
- interlace of two nanofibres, drug and carrier.
- Nanofibre consisting of a blend of the drug and the carrier.

8.2.4 Wound Dressing

In the area of wound and burn treatment, polymer nanofibres show promising potential. Recently, very fine biodegradable polymeric fibres have been directly sprayed (electrospun) onto the skin wounds with the aid of an electrical field to produce a fibrous mat dressing [27–29]. This form of dressing aids the formation of normal skin growth and prevents the formation of scar tissue that results from conventional treatment. Non-woven nanofibrous membranes for wound dressing can be made with pore size ranging from 500 to 1000 nm in order to shield the wound against bacterial penetration by aerosol particle capture.

8.2.5 Haemostatic Devices

In addition to wound dressing, the high surface area per mass of nanofibrous membranes, ranging from 0.25×10^8 to $100 \times 10^8\,\mathrm{mm^2/g}$, is highly efficient for fluid absorption and dermal delivery. As such, polymeric nanofibers are promising in the development of high-efficiency haemostatic devices.

8.2.6 Cosmetics

Current skincare products such as body or facial lotion, ointment, liquid sprays and facial masks may contain dusts, pollen or any form of allergens that may migrate to the more sensitive areas of the body, causing allergies and/or inflammation. Hence polymer nanofibres are being tested as cosmetic skincare masks for the treatment of skin healing or cleansing and other therapeutic purposes with or without various additives [30]. The small interstices and high surface area of nanofibrous skin masks facilitate higher utilization and transfer rate of the additives to the skin for the fullest use of the additive potential. Such cosmetic skin masks made from electrospinning can be gently and painlessly applied to the three-dimensional topography of the skin for healing and skincare.

8.3 Filtration Applications

8.3.1 Filter Media

Though limited in general literature publication, polymeric nanofibres have been used as filter media in industrial filtration over the past two decades [31]. Small fibres in the submicron range, in comparison to larger ones, are well known to provide a higher filter efficiency at equal pressure drop in the interception and inertial impaction stages of the filtration process [32]. A continuing trend is therefore expected when the fibrous diameter goes down to the nanoscale. Nano-fabrication of such filter media looks promising and NonWoven Technologies Inc. of Georgia has developed thin-plate die technology that holds the key to the economic production of meltdown submicron fibres (diameter < 1 μm) for filtration products [33]. In a related development, a US patent has disclosed a method of making a dust filter bag which constitutes a plurality of layers including a carrier material layer and a nanofibre non-woven tissue layer. Nanofibres for applications in pulse-clean cartridges for dust collection and in cabin air filtration of mining vehicles have been discussed [34]. Polymer nanofibres can also be electrostatically charged to modify the ability of electrostatic attraction of particles without increase in pressure drop to further improve filtration efficiency. In this regard, the electro-spinning process has been shown to integrate the spinning and charging of polymers into nanofibres in a single step [35, 36].

8.3.2 Protective Clothing

Military protective clothing serves its purpose in maximizing the survivability, sustainability and combat effectiveness of the individual soldier against extreme weather conditions, ballistics, and nuclear, biological and chemical warfare [37]. Protective clothing for civilians against chemical agents such as sarin, soman, tabun and mustard gas has also become important in recent times. Current protective clothing containing charcoal absorbents has limitations in terms of water perme-ability and the extra weight imposed. As such, it is desirable for protective clothing to possess the following characteristics: lightweight, breathable fabric, permeable to air and water vapour, insoluble in all solvents and highly reactive with nerve gases and other deadly chemical agents. Smith and Reneker [29] invented insoluble linear poly(ethylenimine) nanofibres that have a greater surface area capable of neutralizing chemical agents without impeding the air and water vapour perme-ability to the clothing. A more functional and smart fabric has also been dis-cussed by Nadis [38]. Currently researchers in the US Army Natick Soldier Center are carrying out a project for developing polymer nanofibres for various pro-tective clothing applications [39–41]. Their experimental results revealed that electrospun nanofibres present lower impedance to moisture vapour diffusion and maximum efficiency in trapping aerosol particles as compared to conventional textiles.

8.4 Material Reinforcement

The moduli and fracture resistance have been enhanced in epoxy resin via reinforcement of polybenzimidazole (PBI) nanofibrous non-woven mats with an average diameter of 300 nm [42]. It has also been shown that specific cases of nanofibre reinforcement result in delamination resistance [43], transparent composites [44] and a potential spacecraft application [45]. Carbon nanofibres for composite applications can also be manufactured from their precursors of polymer nanofibres [46, 47]. Compared to commercially available carbon nanofibres produced by a vapour growth technique, converting polymer nanofibres such as polyacrylonitrile (PAN) and mesophase pitch to carbon nanofibres by stabilization and carbonization can produce continuous, uniform, solid carbon nanofibres.

8.5 Electrical Conductors

Polymer nanofibres of varying conductivity can be made by electrospinning conducting polymers such as polyaniline [48–52]. These nanofibres can be used in fabricating small electronic devices such as Schottky junctions, sensors and actuators. Since the rate of electrochemical reaction is directly in proportion to the electrode's surface area, conductive nanofibrous membranes are also suitable for use as porous electrodes in high-performance batteries [49, 53]. It has also been proposed that electrical, ionic and photoelectric conductive nanofibrous membranes have potential application in electrostatic dissipation, corrosion protection, electromagnetic interference shielding and photovoltaic devices [52, 54].

8.6 Optical Applications

Waters *et al.* [55] patented the use of very fine fibres on the submicron scale for liquid crystal optical shutters that can be switched under an electric field between substantially transparent to incident light and substantially opaque. The main part of the liquid crystal device consists of a fibre/liquid crystal composite with a thickness of only a few tens of microns. The choice of fibre size determines the refractive index differences between the liquid crystal material and the fibres, hence it governs the transmissivity of the device. For this type of application, the potential and performance of fine fibres on the nanoscale awaits further experimentation.

8.7 Sensor Devices

Nanofibres produced from polymers with piezoelectric properties (such as polyvinylidene fluoride) produce piezoelectric nanofibrous devices [7]. Polymer nanofibres were also used in developing functional sensors, with the high surface area of the nanofibres facilitating their sensitivity. Poly(lactic-*co*-glycolic) acid (PLGA) nanofibre films were employed as a new sensing interface for developing chemical and biochemical sensor applications [56, 57]. Highly sensitive optical sensors based on fluorescent polymer nanofibre films have recently been reported [58, 59]. Preliminary results indicate that the sensitivities of nanofibre films to detect ferric and mercury ions and a nitro compound (2,4-dinitrotulene, DNT) are two to three orders of magnitude higher than sensitivities obtained from thin film sensors. A single nanofibre coated with two metals at different segments will create a junction, which can be made into a thermocouple to detect inflammation of coronary arteries with extremely fast response times [60]. Such nanothermocouples can be inserted into a cell to monitor the metabolic acticvities at various locations within the cell. Furthermore, multiple nanothermocouples can be circumferentially mounted on a catheter balloon to allow mapping of the arterial wall temperature [61].

8.8 Conclusion

Nanofibres and nanofibre structures are relatively recent materials. A number of publications have appeared in recent years on specific polymeric nanofibers, their processing methods and uses. See Figure 8.7 for a summary of next-generation

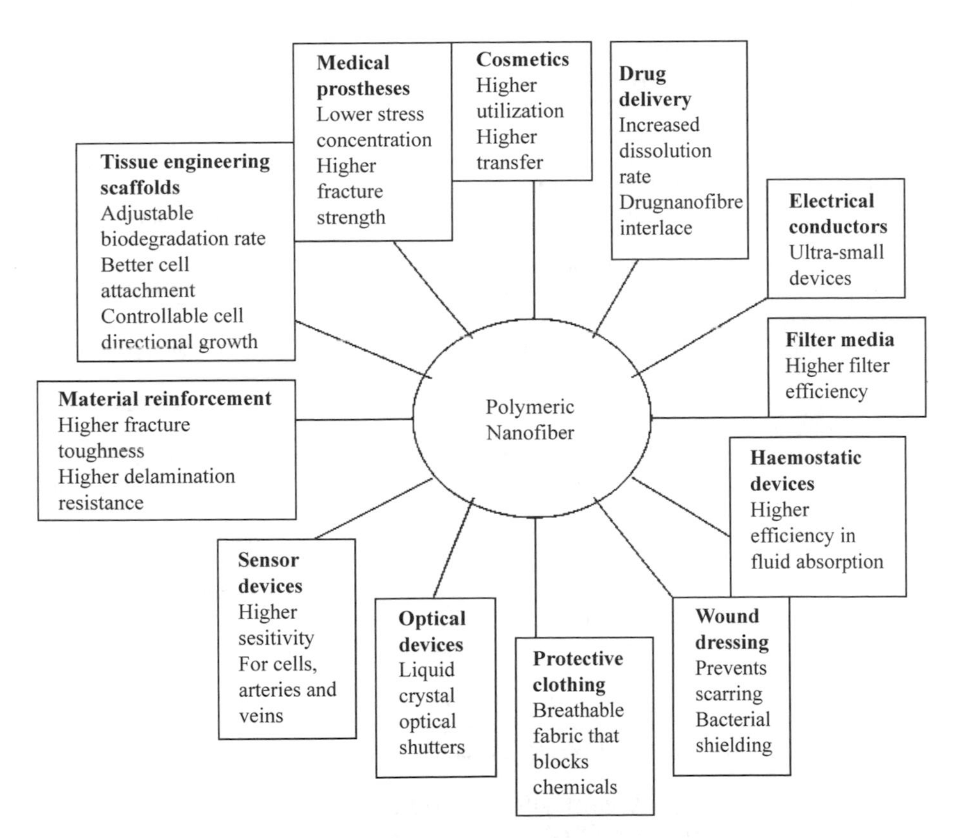

Figure 8.7 Summary and advantages of polymeric nanofibres for next-generation applications

applications. However, there are several areas that require attention for further development of the field. Potential applications of polymer nanofibres have been recognized, but are mainly limited to the laboratory at present. Much greater efforts will be required to commercialize these applications. As a result, research and development of polymer nanofibres will continue to attract the attention of scientists in the near future.

Acknowledgement

The authors acknowledge the efforts by M. Kotaki, R. Inai, C. Y. Xu and F. Yang of the Biomaterials Lab at NUS, and Professor A. Yarin of Technion-Israel Institute of Technology.

References

1. Martin *et al.*, US Patent 4044404, 1977.
2. How T. V., US Patent 4552707, 1985.
3. Bornat A., US Patent 4689186, 1987.
4. Martin *et al.*, US Patent 4878908, 1989.
5. Berry J. P., US Patent 4965110, 1990.
6. Stenoien *et al.*, US Patent 5866217, 1999.
7. Scopelianos A. G., US Patent 5522879, 1996.
8. Athreya S. A. and Martin D. C., *Sensors and Actuators*, **72** (1999) 203.
9. Buchko C. J., Chen L. C., Shen Y. and Martin D. C., *Polymer*, **40** (1999) 7397.
10. Buchko C. J., Slattery M. J., Kozloff K. M. and Martin D. C., *Journal of Materials Research*, **15** (2000) 231.
11. Buchko J., Kozloff K. M. and Martin D. C., *Biomaterials*, **22** (2001) 1289.
12. Yang F., Murugan R., Ramakrishna S., Wang X., Ma Z. W. and Wang S., *Biomaterials*, **25** (2004) 1891.
13. Migliaresi C. and Fambri L., *Macromolecular Symposia*, **123** (1997) 155.
14. Laurencin C. T., Ambrosio A. M. A., Borden M. D. and Cooper J. A. Jr, *Annual Reviews of Biomedical Engineering*, **1** (1999) 19.
15. Huang L., McMillan R. A., Apkarian R. P., Pourdeyhimi B., Conticello V. P. and Chaikof E. L., *Macromolecules*, **33** (2000) 2989.
16. Fertala A., Han W. B. and Ko F. K., *Journals of Biomedical Materials Research*, **57** (2001) 48.
17. Stitzel J. D., Pawlowski K. J., Wnek G. E., Simpson D. G. and Bowlin G. L., *Journal of Biomaterials Applications*, **16** (2001) 22.
18. Boland E. D., Wnek G. E., Simpson D. G., Pawlowski K. J. and Bowlin G. L., *Journals of Macromolecular Science*, **A38** (2001) 1231.
19. Nagapudi K., Brinkman W. T., Leisen J. E., Huang L., McMillan R. A., Apkarian R. P., Conticello V. P. and Chaikof E. L., *Macromolecules*, **35** (2002) 1730.
20. Li W. J., Laurencin C. T., Caterson E. J., Tuan R. S. and Ko F. K., *Journals of Biomedical Materials Research*, **60** (2002) 613.
21. Matthews J. A., Wnek G. E., Simpson D. G. and Bowlin G. L., *Biomacromolecules*, **3** (2002) 232.
22. Xu C. Y., Inai R., Kotaki M. and Ramakrishna S., *Biomaterials*, **25** (2004) 877.
23. Lim T. C., Kotaki M., Yong T. K. J., Yang F., Fujihara K. and Ramakrishna S., *Materials Technology*, **19** (2004) 20.
24. Ignatious F. and Baldoni J. M., PCT/US01/02399, 2001.
25. Kenawy E. R., Bowlin G. L., Mansfield K., Layman J., Simpson D. G., Sanders E. H. and Wnek G. E., *Journal of Controlled Release*, **81** (2002) 57.

26. Zong X., Kim K., Fang D., Ran S., Hsiao B. S. and Chu B., *Polymer*, **43** (2002) 4403.
27. Coffee R. A., PCT/GB97/01968, 1998.
28. Martindale D., *Scientific American*, July 2000, pp. 34–36.
29. Smith D. and Reneker D., PCT/US00/27737, 2001.
30. Smith D., Reneker D., Schreuder-Gibson H., Mello C., Sennett M. and Gibson P., PCT/US00/27776, 2001.
31. Graham K., Ouyang M., Raether T., Grafe T., McDonald B. and Knauf P., Fifteenth Annual Technical Conference and Expo of the American Filtration and Separations Society, Galveston, Texas, 9–12 April 2002.
32. Kosmider K. and Scott J., *Filtration and Separation*, **39** (2002) 20.
33. Ward G. F., *Filtration and Separation*, **38** (2001) 42.
34. Emig D. *et al.*, US Patent 6395046, 2002.
35. Tsai P. P., Gibson H. S. and Gibson P., *Journal of Electrostatics*, **54** (2002) 333.
36. Angadjivand S. A., US Patent 6375886, 2002.
37. Chamberlain G. and Joyce M., *Design News*, 20 August 1990.
38. Nadis S., *Scientific American*, **10** (1999) 74.
39. Gibson P., Schreuder-Gibson H., and Pentheny C., *Journal of Coated Fabrics*, **28** (1998) 63.
40. Gibson P., Schreuder-Gibson H., and Rivin D., *Colloids and Surfaces*, **A187/188** (2001) 469.
41. Schreuder-Gibson H., Gibson H., Senecal K., Sennett M., Walker J., Yeomans W., Ziegler D. and Tsai P. P., *Journal of Advanced Materials*, **34** (2002) 44.
42. Kim J. S. and Reneker D. H., *Polymer Composites*, **20** (1999) 124.
43. Dzenis Y. A. and Reneker D. H., US Patent 6265331, 2001.
44. Bergshoef M. M. and Vancso G. J., *Advanced Materials*, **11** (1999) 1362.
45. Park C., Ounaies Z., Watson K. A., Pawlowski K., Lowther S. E., Connell J. W., Siochi E. J., Harrison J. S. and Clair T. L., *Materials Research Society Symposium Proceedings*, **706** (2002) Z3.30.1.
46. Chun I., Reneker D. H., Fong H., Fang X., Deitzel J., Tan N. B. and Kearns K., *Journal of Advanced Materials*, **31** (1999) 36.
47. Dzenis Y. A. and Wen Y., *Materials Research Society Symposium Proceedings*, **702** (2002) U5.4.1-U5.4.6.
48. Reneker D. H. and Chun I., *Nanotechnology*, **7** (1996) 216.
49. Norris I. D., Shaker M. M., Ko F. K. and Macdiarmid A. G., *Synthetic Metals*, **114** (2000) 109.
50. Macdiarmid A. G., Jones W. E., Norris I. D., Gao J., Johnson A. T., Pinto N. J., Hone J., Han B., Ko F. K., Okuzaki H. and Llaguno M., *Synthetic Metals*, **119** (2001) 27.
51. Ko F. K., Macdiarmid A. G., Norris I. D., Shaker M. and Lec R. M., PCT/US01/00327, 2001.
52. Senecal K., Samuelson L., Sennett M. and Schreuder-Gibson H., US patent application publication, US 2001/0045547.
53. Yun K. S., Cho B. W., Jo S. M., Lee W. I., Park K. Y., Kim H. S., Kim U. S., Ko S. K., Chun S.W. and Choi S.W., PCT/KR00/00501, 2001.
54. Senecal K. J., Ziegler D. P., He J., Mosurkal R., Schreuder-Gibson H. and Samuelson L.A., *Materials Research Society Symposium Proceedings*, **708** (2002) BB9.5.1.
55. Waters C. M., Noakes T. J., Pavery I. and Hitomi C., US Patent 5088807, 1992.
56. Kwoun S. J., Lec R. M., Han B. and Ko F. K., *Proceedings of the IEEE/EIA International Frequency Control Symposium and Exhibition*, 2000, pp. 52–57.
57. Kwoun S. J., Lec R. M., Han B. and Ko F. K., *Proceedings of the IEEE 27th Annual Northeast Bioengineering Conference*, 2001, pp. 9–10.
58. Wang X., Lee S. H., Drew C., Senecal K. J., Kumar J. and Samuelson L. A., *Materials Research Society Symposium Proceedings*, **708** (2002) BB10.44.1.
59. Lee S. H., Ku B. C., Wang X., Samuelson L. A. and Kumar J., *Materials Research Society Symposium Proceedings*, **708** (2002) BB10.45.1.
60. Huang Z. M., Kotaki M. and Ramakrishna S., *Innovation*, **3**(3) (2003) 30.
61. Ramakrishna S., *National University of Singapore Engineering Research Newsletter*, **18**(1) (2003) 4.

9

Nanotechnology Applications in Textiles

David Soane, David Offord and William Ware
Nano-Tex LLC

9.1 Introduction

Nanotechnology is an emerging, highly interdisciplinary field, premised on the ability to manipulate structural materials on the level of individual atoms and molecules. It encompasses expertise spanning over traditional physics, chemistry, material science, computer simulation and electrical and mechanical engineering, yet is not defined exclusively by any one of these disciplines. Many nanotechnology-based innovations have appeared in the literature [1, 2], offering great promise for the future. Indeed nanotechnology has been hailed as the next big thing, which in time could achieve significance comparable to the advent of microelectronics, genomics, wireless communication and the internet. Like all disruptive technologies, the fruits of nanotechnology research and development may require some time to reach maturity. Nevertheless, commercial applications in several consumer-driven industries have begun to emerge.

The underlying efforts responsible for nanotechnology-based advances can be largely divided into two seemingly divergent approaches: precision engineering (top-down) and structure-induced self-assembly (bottom-up). The latter is exactly the focal point of Nano-Tex's efforts. In order to ensure cost-competitiveness and environmentally friendly processes, all Nano-Tex's innovations must employ aqueous solution chemistry, so that high costs associated with vacuum processes

Nanotechnology: Global Strategies, Industry Trends and Applications Edited by J. Schulte
 ISBN: 0-470-85400-6 (HB)

and/or the use of exotic reaction media can be avoided. Furthermore, Nano-Tex's products must be distinct from traditional finishes for textiles in one unique aspect: the reliance on intelligent design and synthesis of starting materials that ultimately lead to surface-induced conformational rearrangements and self-assembly. Most of our ingredients are functional macromolecules that are custom-tailored so that they spontaneously undergo conformational transition, cross-linking and covalent anchoring in the presence of fibre surfaces.

Textile fabrics are one of the best platforms for deploying nanotechnology. Fibres make for optimal substrates where a large surface area is present for a given weight or volume of fabric. The synergy between nanotechnology and the textile industry judiciously exploits this property of large interfacial area and the drastic change of energetics experienced by macromolecules or supramolecular clusters in the vicinity of a fibre when going from a wet state to a dry state. Below we will give a few examples to illustrate the power in realizing the unique opportunities afforded by the intersection of nanotechnology and fabric or fibre treatments.

9.2 Nano-Care and Nano-Pel

In the early days of textile finishing, only simple methods of imparting water repellency to fabrics were available, such as oils, waxes and insoluble soaps. These methods suffered from poor hand (tactile feeling) and insufficient durability to laundering. The durability of the finishes was improved in the 1930s with the introduction of reactive fatty acid water repellents. Silicones, in the 1950s, significantly improved the water repellency performance of the treated fabrics with better hand and durability. It wasn't until the late 1950s and 1960s when fluorochemicals were first used to impart both water and oil repellency, thus achieving stain repellency for the first time. Such a major accomplishment, however, still suffered from low wash-fastness and permanence of observed effects. The root cause was the use of rather generic copolymers (albeit fluoro- and hydrocarbon mixtures). Recently, long-term environmental impact of leachable fluorochemicals has been called into question, making it an urgent issue to ensure minimum usage and attrition. This further supports the strategy of covalent anchoring of nanostructures.

9.2.1 Nano-Whisker Architecture

Nano-Tex has developed two superior water- and oil-repellent products based on custom-designed fluorocarbon-containing polymers: Nano-Pel and Nano-Care. Nano-Pel is a water- and oil-repellent treatment that can be applied to all major apparel fabrics, including cotton, wool, polyester, nylon, rayon and blends. Nano-Care is a product for 100% cotton that imparts wrinkle resistance in addition to water and oil repellency.

Generally, copolymers exhibiting water and oil repellency are comprised of a (meth)acrylate monomer containing a perfluoroalkyl group capable of directly giving water and oil repellency, a fluorine-free monomer capable of improving

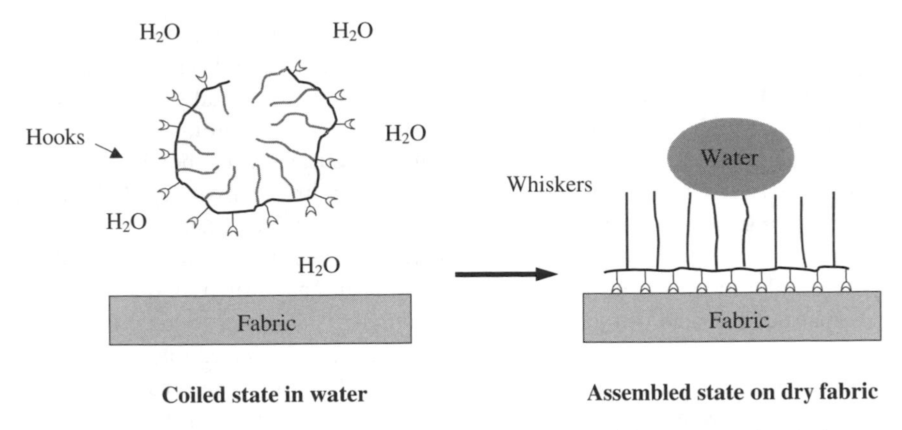

Figure 9.1 Conformational transition of Nano-Pel and Nano-Care chemicals

adhesiveness to fibres, and a monomer capable of ensuring durability through self-cross-linking or reaction with reactive groups on the surface of the materials to be treated. Most commercial copolymers have *N*-methylol groups along the main chain, such as copolymers of perfluoroalkyl-containing (meth)acrylate and *N*-methylol acrylamide copolymers. However, when the fibrous substrate is treated with these copolymers, formaldehyde is produced, which is highly undesirable from an environmental and safety standpoint.

The architecture of the nanowhiskers is depicted in Figure 9.1, where oligomeric or polymeric side branches (brushes) are attached to a flexible spine. Also attached are latent 'hooks' that can form covalent links with functional groups on the fibre surface upon drying and curing. In the aqueous state, the nanostructure coils up to shield the hydrophobic branches within a polar outer layer, as suggested by size information obtained via dynamic light scattering. Upon drying and exposure to heat, the coils unfurl, bringing the polar backbone and multiple hooks in close proximity to the fibre surface (which is generally polar). The brushes project outward from the surface, essentially forming a monomolecular layer to protect against future water or oil intrusion.

Nano-Tex has patented a formulation containing a novel water- and oil-repellent agent capable of binding to fibrous substrates and other materials without the production of formaldehyde. This formulation can impart formaldehyde-free wrinkle resistance and water and oil repellency when combined with a formaldehyde-free resin such as dimethylurea glyoxal (DMUG) or butane tetracarboxylic acid (BTCA).

9.2.2 Polymer Synthesis and Additives

The key ingredient of Nano-Tex's patented water and oil repellents is a copolymer that comprises (a) an agent containing a fluoroaliphatic radical; (b) stearyl (meth)acrylate; (c) a chlorine-containing compound, such as vinylidene chloride, vinyl chloride, 2-chloroethylacrylate or 2-chloroethyl vinyl ether; and (d) a monomer selected from those containing an anhydride functional group or capable of forming an anhydride functional group. This anhydride group can react with various nucleophiles on a fabric surface to form a durable ester bond.

The copolymer may be further copolymerized with (i) hydroxyalkyl (meth)-acrylate to increase the performance and permanency of the resulting copolymer, (ii) a compound such as poly(ethylene glycol) (meth)acrylate to improve solubility of the copolymer in water, and/or (iii) a chain terminator, such as dodecanethiol, mercaptosuccinic acid or other similar compounds, which acts to keep the molecular weight of the polymer low so that it remains readily dispersible in water and can better penetrate the fabric. During the fabric application stage, a catalyst such as sodium hypophosphite is used to induce anhydride formation from the acid-containing monomers in the copolymer. The composition can further comprise other additives such as poly(acrylic acid), which enhances performance and durability of the polymer by some mechanism, possibly by tacking the main ingredient to the surface of the fabric. Other optional additives include an antioxidant such as ethylenediamine tetraacetic acid (EDTA) to reduce substrate yellowing; a permanent softener/extender to improve the hand of the substrate and increase water repellency; a surfactant to emulsify the polymer in water; wetting agents; and/or a plasticizer.

Nano-Pel and Nano-Care impart water and oil repellency to the substrates without adversely affecting other desirable properties of the substrate, such as soft hand (tactile feeling) and breathability. Since their introduction, Nano-Pel and Nano-Care have raised the bar on water- and stain-repellent performance. Since the fluoropolymer is covalently attached to the fibre substrate, we have achieved 100 home laundering durability on 100% cotton substrates.

9.2.3 Process

The application of Nano-Pel and Nano-Care can be accomplished using typical textile mill finishing equipment. The composition can be applied to a fibrous substrate by many continuous finishing methods including dip/pad, spray, foam, knife-coat and kiss roll, followed by drying and curing in an oven. Typically, the dip/pad method is used in which a fabric is immersed in a bath containing the composition followed by passing the fabric through two rollers that squeeze out excess solution. The treated substrate is then dried and cured to allow reaction of the polymer with the textile and with itself. One key step to ensure performance durability is to start out with a clean substrate. Since the durability depends directly on the covalent attachment of the polymers to the fabric substrate, it is imperative that the surface is not blocked by sizes, oils or contaminants. Therefore, substrates must receive a vigorous scour before to the application process.

9.2.4 *Testing and Performance Criteria*

The performance of water- and oil-repellent fabrics is tested by two methods: spray rating and oil rating. The spray rating (SR) of a treated substrate is a value indicative of the dynamic repellency of the treated substrate to water that impinges on the surface, such as encountered by apparel in a rainstorm. The rating is measured by Standard Test Number 22, published in the 1977 Technical Manual and Yearbook of the American Association of Textile Chemists and Colorists (AATCC), and is expressed in terms of the spray rating of the tested substrate. The spray rating is obtained by spraying water on the substrate and is measured using a 0 to 100 scale where 100 is the highest rating.

The oil repellency (OR) of a treated substrate is measured by the American Association of Textile Chemists and Colorists (AATCC) Standard Test Method No. 118-1983, which is based on the resistance of a treated substrate to penetration by oils of varying surface tensions. Treated substrates resistant only to Nujol (mineral oil), the least penetrating of the test oils, are given a rating of 1, whereas treated substrates resistant to heptane (the most penetrating of the test oils) are given a rating of 8. Other intermediate values are determined by testing with other pure oils or mixtures of oils, as shown in Table 9.1.

Table 9.1 Standard test liquids: AATCC oil repellency rating number and composition

Rating number	Composition
1	Nujol
2	Nujol/n-hexadecane 65/35
3	n-Hexadecane
4	n-Tetradecane
5	n-Dodecane
6	n-Decane
7	n-Octane
8	n-Heptane

The durability of the finish is assessed through laundering under normal service conditions in home laundry and drying machines with a common detergent such as Tide. To control the quality of goods made at a given mill, the finished goods must pass certain water and oil repellency requirements after a given number of home launderings.

9.2.5 *Future Directions in Repellency*

Nano-Tex is continually improving its water- and oil-repellent copolymer as well as the other components in the formulation. It is continuing optimization of the fabric preparation and application methods. And it is continuing to research new products that add additional benefits to the finished products.

Nano-Tex is currently researching methods to increase the strength of cotton fabrics after the application of a wrinkle-free resin, as in the Nano-Care formulation. Typically, resin finishes decrease the tensile, tear and abrasion strengths of the cotton being treated by two mechanisms. First, the resin covalently cross-links the cotton, thereby making it more brittle. Second, the cross-linking reaction itself is carried out under an acidic pH. Together with elevated temperature needed to cure the treated fabrics, depolymerization of the cotton cellulose occurs. This second mechanism can be mitigated and we have indeed achieved appreciable gains in tensile and tear strengths and even more noticeable gains in abrasion resistance.

Although Nano-Pel and Nano-Care repel stains, under pressure it is possible for a stain to penetrate the barrier and become embedded in the fibre. Nano-Tex is researching a means not only to impart water and oil repellency to a substrate, but also to enable facile release of an embedded stain once it is immersed in an aqueous soap solution. The approach is a combined use of fluorinated and hydrophilic components. When exposed to the air (which has a very low surface tension), the low-surface-tension fluorinated component orients itself toward the air, thus providing oil and water repellency. However, when the substrate is immersed in soapy water, the hydrophilic component orients itself such that it is exposed to the water, allowing the soap to penetrate the fibre and remove stains. This mechanism has been shown to give stain-release products that are simultaneously water- and oil-repellent with both attributes durable up to 30 home launderings.

9.3 Nano-Dry

A treatment that builds a three-dimensional molecular network surrounding a fibre (i.e. the Nano-Net architecture) is called Nano-Dry. This hydrophilic, or moisture-loving, treatment is applied to polyester and nylon fabrics (Figure 9.2).

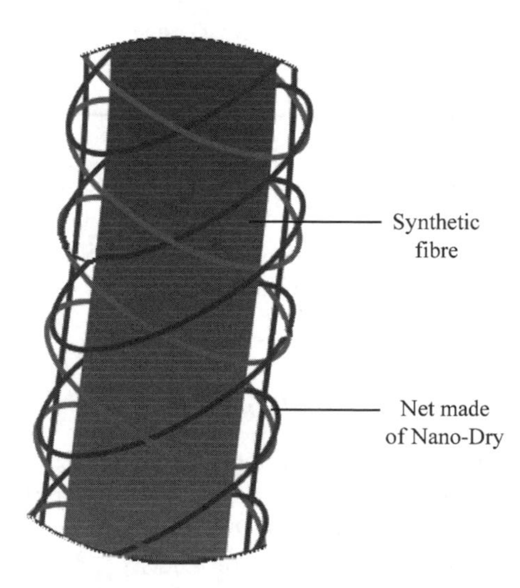

Figure 9.2 Three-dimensional molecular 'net' of Nano-Dry

The treatment durably attaches a hydrophilic network to a hydrophobic substrate without altering the other properties of the material, such as strength, colour fastness and hand (tactile feel).

Synthetic textile materials, such as nylon and polyester, are uncomfortable to wear due to their poor permeability to water. In hot weather, sweat cannot easily penetrate (or wick) through these fabrics and evaporate. The poor wicking and permeability are due to the natural hydrophobicity of nylon and polyester polymers; water does not readily spread out over surfaces composed of these materials. Nylon and polyester also often exhibit static cling and stain retention due to their hydrophobicity.

It is therefore desirable to find a way of imparting durable hydrophilic properties to nylon, polyester and other synthetic materials. This may be achieved by attaching hydrophilic materials to the hydrophobic fibres. Imparting hydrophilic properties to the hydrophobic substrate will also diminish or eliminate static cling and enable the release of stains during laundering.

9.3.1 Basic Approach

Typical treatments for synthetic fabrics to impart hydrophilic properties rely on film formation on the surface of the fabric and self-cross-linking properties of the chemical to achieve some level of durability to laundering. However, these treatments suffer from the conflicting refinements of moisture absorbtivity and adhesion to a synthetic and inherently hydrophobic substrate. An unattached hydrophilic film will absorb moisture in the wash, and upon abrasion it will tend to deteriorate and fall off after repeated washings. The more hydrophilic the film, the less durable the resulting film. Conversely, the higher the durability, the less hydrophilic character the film will have. As a consequence, traditional commercially available hydrophilic treatments for polyester and nylon have durability only up to 5–10 home launderings. One approach employed to overcome the lack of durability is to increase the amount of chemical deposited on the surface. However, this technique quickly reaches diminishing returns on performance as the excess chemical begins to affect the hand of the fabric, it may affect the colour shade, and it will increase the cost.

The Nano-Dry treatment achieves its durability not by film formation, but by the combination of covalent attachment to the fibre surface and the use of nanomolecules. Covalent bonding to the fibre surface allows attachment of large, highly hydrophilic, superabsorbent materials. Synthetic surfaces (e.g. polyester) often provide few reactive sites for chemical attachment. However, a large molecule that covers a significant portion of the fibre surface needs only a few attachment points to anchor durably. Large molecules are generally difficult to work with in textile processing due to their high viscosities in aqueous solutions. However, specific process conditions can be optimized so the molecules arrange themselves in a compacted conformation that minimizes viscosity. This, along with a clean fibre substrate and significant heat to help catalyse the attachment chemistry, results in a very efficient, durable treatment. A typical solids loading for Nano-Dry treatment onto the fabric surface is around 0.1–0.15% by weight of the fabric, compared to 0.8–4.0% by weight of fabric for conventional film-forming hydrophilic treatments.

Figure 9.3 contrasts the difference between an untreated fabric and a Nano-Dry treated fabric. The distinction is obvious. The performance of water absorbency of the treated fabric is durable for the life of the garment on both polyester and nylon substrates.

Figure 9.3 The phenomenon of water droplet spreading on a Nano-Dry surface (right) as compared to a non-treated control sample (left)

9.4 Nano-Touch

Use of blended textiles has decreased over recent years in favour of 100% cotton fabrics that offer good appearance and comfort. However, the use of 100% cotton yarn and fabrics has its disadvantages. Primarily, these fabrics tend to shrink and wrinkle. The most popular method of controlling cotton shrinkage and wrinkling for apparel outerwear is to react the cotton fabric with cross-linking resins, which often emit formaldehyde upon cure. Formaldehyde is considered to be a hazardous chemical. It is dangerous to handle during processing and poses a health risk to the ultimate consumers. Additionally, formaldehyde-based resins, when used to control the shrinkage and wrinkling of cotton or cotton-blend fabrics, significantly degrade the abrasion resistance and strength properties of the fabric, thus making them more prone to fabric holes and scuffing. Although non-formaldehyde resins have been invented (e.g. polycarboxylic acids), they do not work as well as formaldehyde-emitting resins, they are more expensive, and they still suffer from the same fabric strength loss.

The use of pre-washing to control shrinkage is also less than satisfactory, because it wastes energy and it gives new garments a worn appearance. Mechanical compaction has also been used to control the shrinkage of cotton fabrics. However, this process is expensive because of the high working loss. It is also not a permanent solution, as compacted garments tend to return to their precompacted dimensions. Neither of these methods addresses the wrinkling tendency of cotton. For these reasons, resin treatment remains the preferred method to control the shrinkage and wrinkling of cotton fabrics, despite their many drawbacks.

With the advent of synthetic textile fibres, the possibility arose for producing continuous filament yarns with greater strength and more durability than those formed of staple fibres. Furthermore, synthetics tend not to have wrinkling or shrinkage problems. Products made from synthetic yarn have excellent strength properties, dimensional stability and good colour fastness to washing, dry-cleaning and light exposure. The use of 100% polyester knit and woven fabrics became extremely popular during the late 1960s and through the 1970s. More recently, continuous filament polyester fibre has also been cut into staple that can then be spun into 100% staple yarns or blended with cotton or other natural fibres. However, 100% polyester and polyester-blended yarns and fabric made from these yarns have a shiny and synthetic appearance, they are clammy and prone to static build-up in low humidity, and they tend to be hot and sticky in high-humidity conditions. Additionally, because of its high tensile strength, polyester fibre is prone to pilling in staple form and picking in continuous filament form.

Several attempts have been made to produce fabrics with the positive qualities of both cotton and polyester fibres and without their negative attributes. These attempts have included blending, sheath/core yarn spinning, and sheath/core fibre composites (grafting). These methods require modifications on the fibre scale, but were not possible on the fabric scale until the current invention by Nano-Tex.

9.4.1 Overview of Composite Fabrics

Yarns have been manufactured for many years with a distinct sheath/core configuration. For example, literature exists for ring spun yarns having synthetic fibres in the core and a cotton sheath (US Patents 4,711,079, 5,497,608, 5,568,719 and 5,618,479). A well-known method of spinning homogeneous and composite yarns has been ring spinning, which has several advantages. Ring spinning produces a strong yarn of high quality, at a low capital investment per spindle. Unfortunately, ring spinning is a comparatively slow process, capable of producing only about 10–25 m of yarn per minute, which greatly increases the cost of the final product. In addition, it is impossible to avoid a highly fluctuating sheath content, resulting in a yarn of core/sheath filaments including that include sections without any sheath content. Nevertheless, since no other previously known process could produce the strength or feel of ring spun yarn, this process is still used when the demand for its strength and feel justifies the high costs involved.

The concept of producing sheath/core fibre composites by grafting is relatively new (US Patents 3,824,146, 5,009,954, 5,272,005, 5,387,383). It involves a fibre-coating process in which a synthetic core is passed through a fibre-coating die, where it contacts viscose rayon. The rayon coating is then regenerated in a sulphuric acid bath. The core fibre dominates the mechanical properties of the composite fibre, and the rayon skin dominates the surface properties. Typically, the skin does not adhere well to the core, particularly if the core has a smooth surface. Many adhesion promoters have been suggested, with varying degrees of success. Unfortunately, the long reaction times make this method expensive and slow.

9.4.2 Basic Approach

Nano-Tex has launched a project to develop fabrics that exhibit the positive qualities of cotton and synthetics but without their negative qualities. The resulting process is fast, economical and transparent to current textile manufacturing practices, such as sanding, weaving and dyeing.

The Nano-Touch treatment has been developed to create a permanently attached carbohydrate sheath (carbohydrates have the desired properties of cotton, which is itself considered a carbohydrate) around each synthetic fibre of the web. This treatment endows the treated web with the most desirable characteristics of the synthetic core and most desirable characteristics of the natural, carbohydrate sheath.

9.4.3 Process

The polyester or nylon core material, in fabric form, is passed through a bath containing an aqueous solution of water-soluble carbohydrate and cross-linker and, if necessary, a suitable cross-linker catalyst. The fabric is padded to remove excess liquor and heated to dryness to form a thin film of carbohydrate on the surface of the synthetic core. The fabric is then cured at a temperature high enough to cause reaction between the cross-linker, synthetic core and carbohydrate to form covalent bonds between the core material and the carbohydrate. Simultaneously, cross-links are formed between the carbohydrate molecules themselves, forming the sheath. Figure 9.4 is a schematic diagram of the resulting architecture and Figure 9.5 is a TEM cross section of Nano-Touch treatment around a polyester fibre.

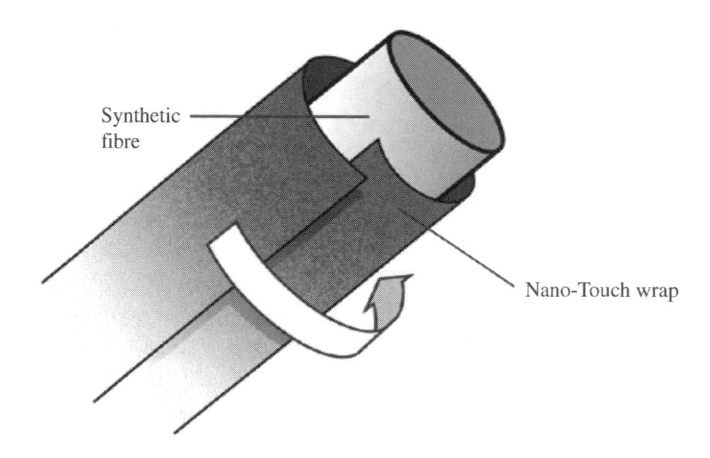

Figure 9.4 Schematic of a Nano-Touch treated fibre

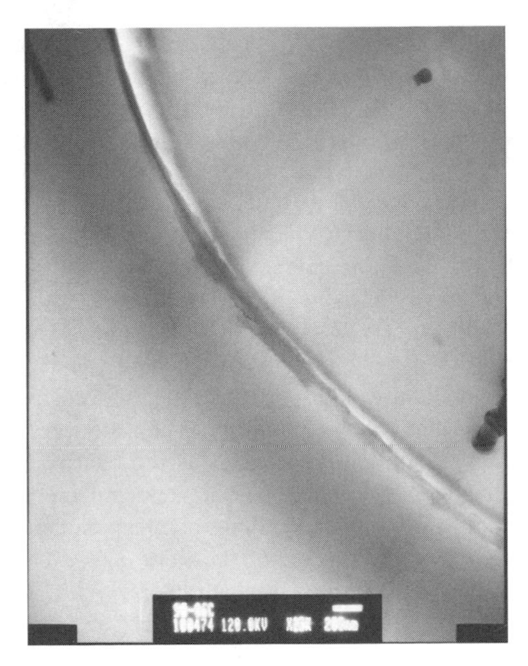

Figure 9.5 TEM cross section of Nano-Touch treatment around a polyester fibre, showing a sheath layer thickness in the range 100–250 nm

9.4.4 Performance Characteristics

By the nature of the wrap architecture, the enhanced properties of the finished fabric are surface related. Surface-related attributes such as feel (hand), moisture absorbency, matte finish and reduced static build-up are enhanced and the best properties of the synthetic fibre are retained at the core level, including strength, colour fastness and crease retention.

Nano-Touch treated fabric regains four times the moisture content compared to normal polyester. Moisture regain is a measure of the ability of a fabric to absorb moisture from, or release moisture to, the surrounding air until an equilibrium moisture content is attained. For polyester fabrics, moisture regain is very low, typically around 0.3% at 70 °F (21 °C) and 65% relative humidity. Under the same conditions, Nano-Touch treated polyester has a 1.2% moisture regain. The increased moisture regain improves the 'natural' feel of the fabric.

Another attribute of the carbohydrate sheath wrapping the fibre is improved moisture wickability. Since the wrap layer has been designed to be 'cotton-like', the sheath is naturally hydrophilic and will wick moisture in a similar manner to the Nano-Dry treatment.

Materials that are non-conductors of electricity are known as dielectric materials. Whenever charges build up on their surfaces, the charges will persist until some pathway forms to conduct them away. As the insulative property of a dielectric material increases, the charge potential can also increase, resulting in problems associated with static electricity. Almost all the fibre materials can be classified as dielectric materials. This is especially true of synthetic fibres, e.g. nylon, polyester, polyolefins, and acrylics. Natural fibres such as cotton and wool can also have problems associated with static charge build-up. However, this problem tends only to occur at low relative humidities (typically less than 35%). In contrast, synthetic fibres will have static build-up problems even at much higher relative humidities [3].

Increasing the conductivity of the textile fibre is the key to increasing the rate of static dissipation. A wide variety of anti-static finishes have been developed; all of them work by increasing the conductivity of the fibre. Nearly all these materials rely on water as the medium to transport the charged species and therefore their usefulness depends on atmospheric humidity. They are very effective in moist atmospheres, but their effectiveness decreases as the relative humidity decreases. Common reagents include hydrophilic surfactants, polyelectrolytes, long-chain quaternary ammonium salts, polyethoxylated polymers and any other hygroscopic material that can be left on the surface of the fibre. However, these reagents are not permanent since they will wash or wear away.

As described earlier, the Nano-Wrap architecture creates a permanent attachment of the carbohydrate sheath (Nano-Touch). Additionally, the hydrophilic nature of the wrap, the increased moisture regain, and the fact that this additional moisture content is localized to the surface of the fibre all cause Nano-Touch treatment to exhibit durable anti-static performance.

The lower the resistivity of a surface, the better its ability to conduct a charge and dissipate static build-up. Figure 9.6 compares the measured surface resistivity of polyester fabric treated with Nano-Touch to control samples of untreated polyester and a natural cotton fabric.

The data shows that the Nano-Touch fabric has less resistivity (is more conductive) than the same polyester without the carbohydrate wrap architecture. Also, after

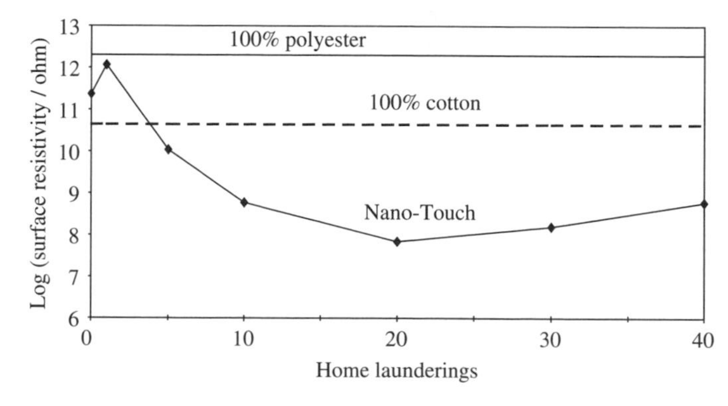

Figure 9.6 Surface sensitivity of Nano-Touch treated fabric compared to 100% cotton and 100% polyester at 50% relative humidity and 73 °F (23 °C)

a few home laundering cycles, its anti-static performance improves and actually becomes significantly better than a similarly weighted cotton fabric.

9.5 Future Trends

The application of a carbohydrate encapsulation layer to a fabric also offers the opportunity to simultaneously finish the fabric with auxiliary components that do not have the innate ability to bind durably to the fabric. In this way, the carbohydrate sheathing acts as a binder to impart durability to non-substantive auxiliary components that are co-applied with the sheathing finish. Alternatively, the auxiliary may have substantivity to the carbohydrate finish and can be applied in processing after application of the carbohydrate encapsulation layer. In either method, the base fabric is endowed with a number of properties that cannot be attained without the use of the encapsulation layer.

Some examples of auxiliary components include infrared-absorbing compounds such as carbon black, a chitin resin or other known infrared-absorbing pigments that can be permanently incorporated onto the fabric to minimize detection from night vision equipment. Generally, the infrared region of interest lies in the range 1000–1200 nm. Fabric treated with an encapsulating layer containing infrared-absorbing materials would have infrared absorptive ability as well as other beneficial properties belonging to the encapsulating layer; such fabric may be of particular interest in military applications.

Similarly, ultraviolet-blocking compounds can be incorporated to protect the fabric material itself or the wearer of the garment from ultraviolet rays. Coloured pigments or dyes may be incorporated in the outer layer. Magnetic colloids can be used in the sheath to provide data storage capabilities to the fabric. Bioactive agents (e.g. insect repellents, antimicrobials and pharmaceuticals) are also examples. Odor-absorbing compounds and neutralizers (e.g. activated charcoal or cyclodextrins) or, alternatively, a fragrance that one wishes to release in a prolonged fashion by using, for example, hydrolysable linkers, may be applied this way. Flame-retardant chemicals and additional anti-static agents may be incorporated into the sheath as well. Combination of Nano-Wrap with novel chemistry, when applied to textiles, leads to many existing opportunities. The future is indeed unbounded.

References

1. Scientific American (2002) *Understanding Nanotechnology*, Warner Books, New York.
2. M. Gross (1999) *Travels to the Nano World*, Plenum, New York.
3. S. Adanur, (1995) *Wellington Sears Handbook of Industrial Textiles*, Technomic Publishing, Lancaster PA.

10

Measurement Standards for Nanometrology

Isao Kojima and Tetsuya Baba
Metrology Institute of Japan

Scales as measurement standards have been indispensable to humankind since ancient times. Today, as economic activity is becoming more and more global and large amounts of materials, parts, products and services are being distributed across national boundaries, measuring standards are becoming established as a vital element in the infrastructure of global economic activity. With regard to the industrial applications of nanotechnology it is anticipated that measuring standards will occupy a more important position than ever in the future since nanotechnology is expected to be the most promising driving force for the development of future industrial science and technology in various fields. Novel fabrication processes based on nanotechnology will become really powerful when the processes are reproducible and reliable. Standard materials and metrology developed for use in nanoscale characterizations are the key tools for establishing reliability in information technology, environmental technology and biotechnology, all of which are now under research and development based on using nanotechnology.

The concept of the traceability of measurement results assumes that the measurand is independent of the method of measurement and, by its very nature, is to be ensured practically only by coordination on various measurement and analytical methods based on scientifically accepted studies. This coordination involves the establishment of measurement standards, reference materials, technical

Nanotechnology: Global Strategies, Industry Trends and Applications Edited by J. Schulte
 ISBN: 0-470-85400-6 (HB)

requirements for calibration, and field studies for the uniformity, stability and reproducibility of nanometrology.

The National Metrology Institute of Japan (NMIZ) in the Institute of Advanced Industrial Science and Technology (AIST) is engaged in R&D for the nanometrology system ensuring the reliability of the product of national scientific and industrial activities under the auspices of the governmental nanotechnology programme. Its R&D spectrum covers the establishment of national calibration technology for various physical properties, including dimensional properties of the reference materials and a database for practical application and elaboration of the technical requirement for the practical calibration. It covers measurands such as line width, layer thickness, size and physical properties of nanoparticles and nanopores, a surface analysis database and the thermal properties of nanomaterials and nano-interfaces.

Two national projects are currently working on these assignments with support from the Japanese Ministry of Economy, Trade and Industry (METI) and the New Energy and Industrial Technology Development Organization (NEDO).

- The R&D of Three-Dimensional Nanoscale Certified Reference Materials Project aims to develop scales for lateral and depth directions. The research period is from 2002 to 2006.
- Nanomaterial Process Technology/Nanotechnology Material Metrology Project aims to develop measurement methods and related reference materials for nanotechnology. The research period is from 2001 to 2007.

The following sections give more details on the R&D they are doing.

10.1 R&D of Three-Dimensional Nanoscale Certified Reference Materials Project

This project is developing the nanometric 'scale' that can be applied to the technologies to control, process and measure the nanostructure with superfine precision.

The nanoscale is intended to offer the calibration traceable to the national measurement standards for lateral and depth directions, and to serve as the certified reference materials (CRMs) to which the certification value and uncertainty are assigned. NMIJ/AIST is presently supplying the measurement standards for a one-dimensional grating scale (0.2 to 8 μm) and GaAl/AlAs superlattice reference material (Figure 10.1). The scanning electron microscope (SEM), which is integrated with the calibrated standard microscale (240 nm pitch) of Figure 10.1(a) is marketed as the CD-SEM for semiconductors. It is one of the major measurement tools developed in Japan that contributes to the field of semiconductors.

In this project, the aim is to develop a nanoscale of ~25–100 nm pitch for the lateral direction, and of ~3–10 nm thickness for the depth direction. These targets were decided by referring to the International Technology Roadmap for Semiconductors (ITRS), which predicted, for example, that the so-called technology nodes by the half-pitch of devices would be 45 nm and 22 nm in 2010 and 2016, respectively [1]. Figure 10.2 shows the outline of the project.

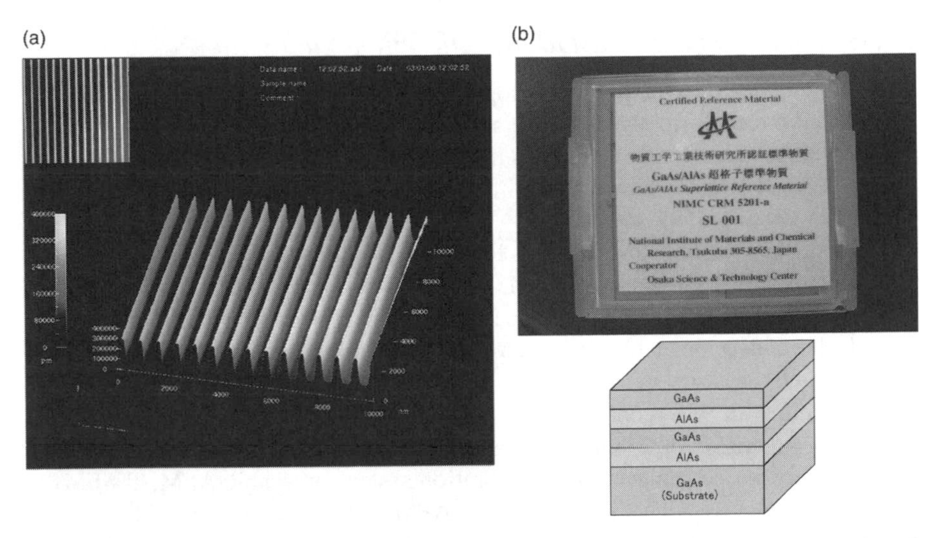

Figure 10.1 The one-dimensional grating scale is an important measuring tool developed in Japan

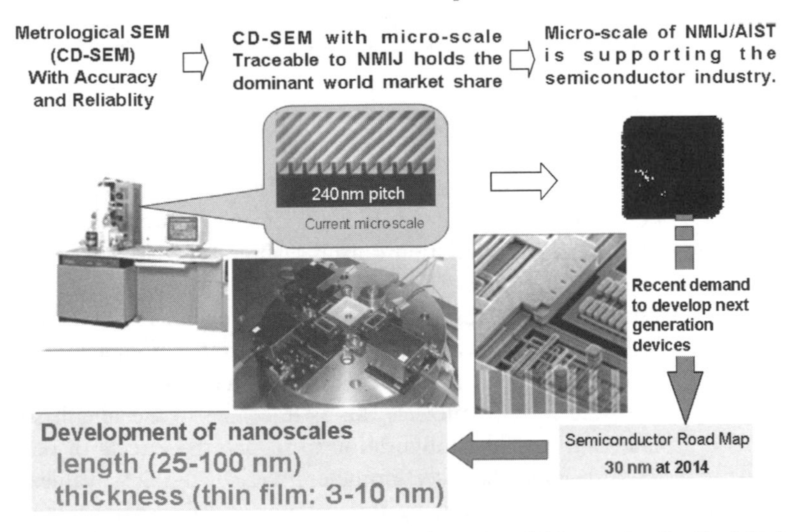

Figure 10.2 R&D overview for the Three-Dimensional Nanoscale Certified Reference Materials Project

10.1.1 Development of a Lateral Direction Nanoscale

This theme aims to develop a technology and calibration system for producing the nanoscales traceable to the length standards provided by the wavelength of an iodine-stabilized HeNe laser. It is also intended that the scales will be supplied after calibration as CRMs in accordance with lateral nanoscales that have the shape of a one-dimensional grating structure.

One of the most promising methods for the calibration of nanometre length scales is the atomic force microscope (AFM) that has a resolution at the atomic level. A prototype system equipped with laser interferometers on the *X*, *Y* and *Z* axes has already been developed [2]. An example of the measurements is given for the 240 nm pitch microscale in Figure 10.1(a). The uncertainty estimated for the scale was 0.17 nm with 95% confidence. However, in order to calibrate a scale with a minimum graduation of 25 nm, it is required to make the overall uncertainty even smaller. This can be achieved by a calibration system (traceable AFM, T-AFM) as shown in Figure 10.3. This is basically an AFM system equipped with laser interferometers having a resolution of about one-fifth the size of an atom, and can be traceable to a length standard.

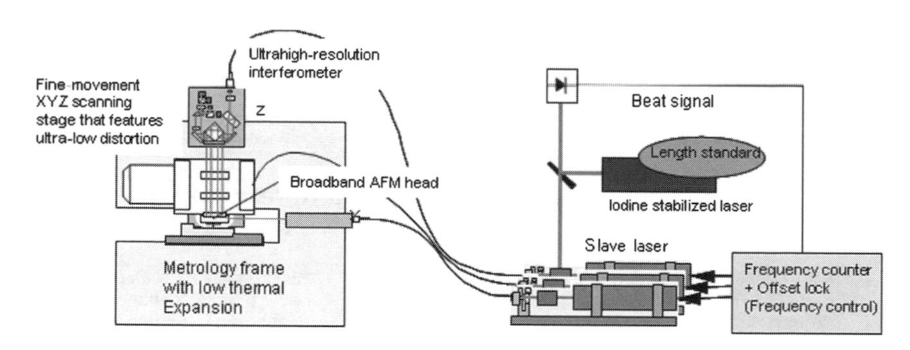

Figure 10.3 Concept of the traceable AFM

For the new measurement system, firstly an *XYZ* fine movement mechanism is required for featuring angular variation during scanning along each optical axis to a few tens of nanoradians (nrad). Secondly, the fine movement mechanism is mounted on a metrology frame featuring low thermal expansion to achieve the three-dimensional displacement with homodyne laser interferometers having a resolution of 0.02 nm. Thirdly, an iodine-stabilized offset lock laser is employed as the light source of the interferometers; this is traceable to the length standard and adopts a symmetrical optical configuration to reduce the effects of deadpath. Finally, we expect to develop the interferometers with uncertainty values below 0.1 nm by building a system for reducing the cycle error. Furthermore, the broadband AFM head employs an AC mode cantilever and the AFM probe is in the form of a sharp-pointed probe.

The one-dimensional diffraction grating that has the size required as a CRM is fabricated by selecting the optimum nanofabrication technology from the technologies that have been established in the semiconductor industry, such as superlattice film fabrication technology, electron beam lithography and X-ray lithography. Several kinds of lateral nanoscales, with 100 to 25 nm per graduation depending on the substances, are under test by taking into cosideration (1) the user's convenience (so that the nanoscales can be replaced whenever damaged); (2) association with industry; and (3) the necessity of proving safety measures.

The nanoscales under development will have 10 000 to 40 000 graduations per 1 mm distance. Each one of these many graduations contains small errors from the nominal value or variances due to imperfections in the fabrication process. This inevitable phenomenon makes it essential to calibrate the small errors and variations of graduations using an accurate length-scale calibration system.

10.1.2 Development of a Depth Direction Nanoscale

A depth direction scale is required to quantify the properties such as the film thickness and the depth of injected impurities, for example in the pn junction layer of the MOS field-effect transistor (MOSFET). Since most of the practical methods to measure the depth distribution depend on material properties, unlike the in-plane direction nanoscales, the depth direction nanoscales necessitate the control of a wide range of factors besides the length (film thickness), such as the density of the films, the uniformity of their compositions in a depth direction, the roughness of the surface and interface, and the specific structures of the boundaries like a transition layer. Considering the needs and the wide range of applications that have already been achieved in the semiconductor field, we have set the objective of developing CRMs for use in the depth direction scale calibration of GaAs/AlAs superlattice and ultra-thin SiO_2 film on Si.

With GaAs/AlAs we have already fabricated and supplied standard materials with a 25 nm film thickness (Figure 10.1) and are now targeting improvements to the 10 nm level. On the other hand, for the SiO_2/Si ultra-thin films, the thickness of the surface contamination layer [3, 4] and the transition layer will affect the margin of error in the measurements. It is thought that if the film is fabricated using thermal oxidation by oxygen molecules, structural transition layers may be produced at the boundaries that depend on the fabrication conditions. These layers will pose a serious problem, particularly for the development of ultra-thin film reference materials. Therefore, in place of using the thermal oxidation technique, we plan to apply the ozone (O_3) oxidation technique. Ozone has a higher oxidizing activity than oxygen.

AIST has already developed and established a technology for the safe generation and control of 100% concentration ozone gas [5] and has also succeeded in low temperatures fabrication of a high-quality SiO_2 film on Si substrate, although the size has been limited to 10 mm × 10 mm [6]. It has been confirmed that the thickness of the structural transition layer on the oxide film is extremely small

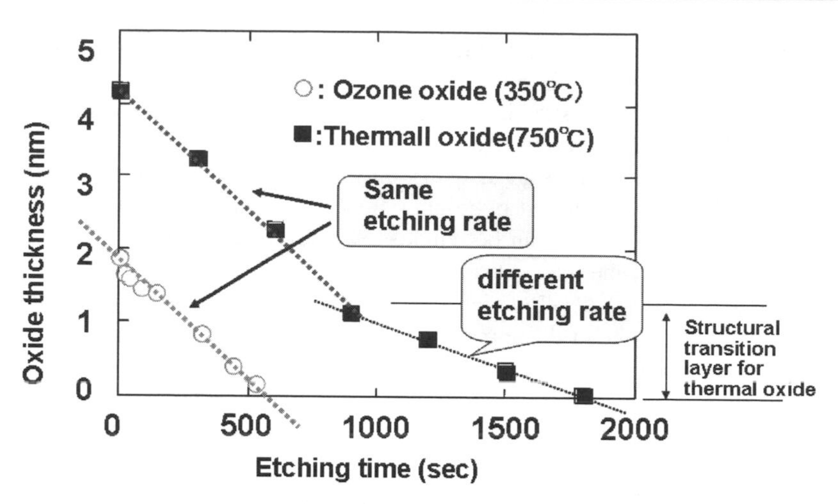

Figure 10.4 Comparison of the boundary structural transition layer between a thermally oxidized film and an ozone-oxidized film

[7]. Figure 10.4 shows the results from measuring the thickness of the structural transition layer using a chemical etching technique. It has actually been shown that the structural transition layer is extremely thin compared to the thermally oxidized film. Based on this achievement, we are developing a technology for fabricating samples of a size suitable for CRMs by setting it as our primary objective.

It is also essential to use a highly accurate thickness measurement method. Here we plan to develop the X-ray reflectivity technique (traceable XRR technique) as a film thickness determination method that is traceable to the higher standards. When the angle of incidence of X-rays into a measurement sample exceeds a critical angle, their reflectivity suddenly decreases with increasing incidence angle and an oscillation structure appears, called the Kiessig fringe. As the oscillation period is strongly related to the film thickness, the thickness can be determined by observing the oscillation structure while precisely controlling the incidence angle. To ensure traceability, it is necessary to determine the X-ray incidence angle and incident X-ray wavelength. Figure 10.5 shows the configuration scheme for the XRR system.

Since the oscillation period increases as the film thickness decreases, the technique requires a high-intensity X-ray source especially for the ultra-thin films. For this purpose, the system uses an X-ray generator having an 18 kW output from a rotating Cu target together with X-ray condensing optics. The scattering angle 2θ can be measured accurately using a high-resolution goniometer. The goniometer is controlled with an angle calibrator that is traceable to the national angle standard, and the error in the angle measurement is reduced to below 1 arcsecond. This development will make it possible to implement a highly accurate film thickness calibration with an accuracy of less than one atomic layer.

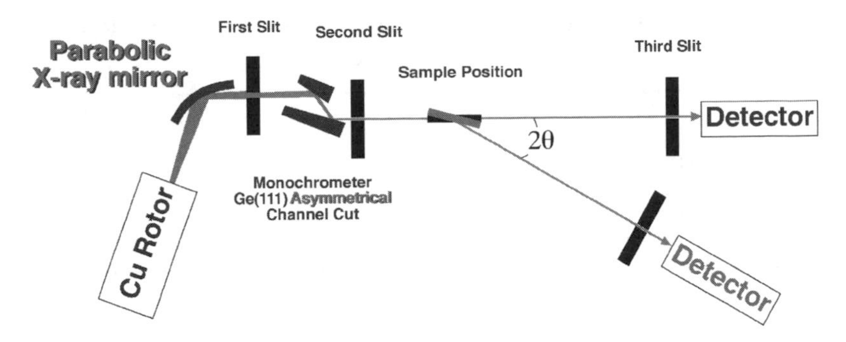

Figure 10.5 Configuration of a traceable XRR system

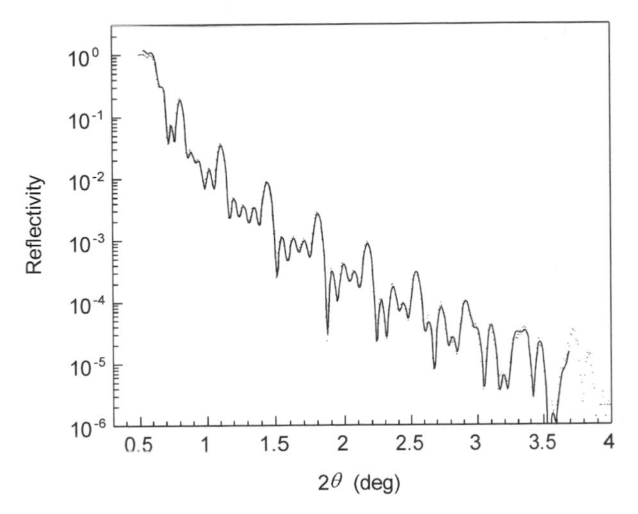

Figure 10.6 Non-linear least-squares fitting of the X-ray reflectivity profile for a GaAs/AlAs superlattice

Figure 10.6 shows an example of the XRR measurement for the GaAs/AlAs superlattice CRM (NIMC CRM5201-a). Least-squares fitting revealed properties such as thickness, density, surface roughness and interface roughness for all four layers (Table 10.1). The repeatability of the thickness measurement was better than 0.5% except for the thickness of the top surface layer, because it increased slightly with repeated measurements. Thus, the uncertainties were about 0.3 nm with 95% confidence, the smallest among multilayer CRMs supplied in the world.

10.1.3 International Comparisons of Nanometric Scales at BIPM

The supply of certified standard reference materials that feature absolute values of length and thickness would be meaningless if their values were based exclusively

Table 10.1 Evaluated properties of the GaAs/AlAs superlattice CRM

	$\delta/10^{-6}$	$\beta/10^{-6}$	Thickness (nm)	Roughness (nm)
Oxide	8.132	0.266	1.241	0.361
GaAs	14.535	0.421	23.385	0.457
AlAs	10.709	0.296	22.572	0.334
GaAs	14.497	0.421	23.313	0.323
AlAs	10.581	0.296	22.589	0.361
Substrate	14.458	0.421	10 000	0.349

on standards specific to Japan and isolated from other world standards. Construction of a traceable system should consider international traceability. Thus, under the leadership of the Bureau International des Poids et Mesures (BIPM) an international comparison of various quantities is attempted in order to acquire an acceptable international uniformity [8]. Some of the nanoscales developed in the framework of this project have already been subjected to preliminary international comparisons.

For example, in 2000 a one-dimensional grating that had pitches of about 300 and 700 nm was subjected to a supplementary comparison by the Consultative Committee for Length (CCL) [9]. Various national metrology institutes (NMIs) joined in the comparison and calibrated according to their own primary national length standards for nanometrology. The calibrations were made using optical diffraction (OD), optical microscopy (OM) and scanning probe methods (SPM). Each calibration result was reported with its claimed uncertainty, which was deduced from intensive evaluations on the various sources of uncertainty.

Uncertainties in the wavelength of the laser applied to the OD, in collimation of the laser beam, in alignment of the laser beam with the optical axis of the grating, in the measurement of the diffraction angle, in the non-uniformity of the pitch over the grating, etc., had to be evaluated carefully and reported. However, since the pitch is a macroscopic measurand, the periodicity of the line pattern may differ from line to line and between both ends of a line. On the other hand, since SPM is a microscopic tool, it is capable of appreciating local deviations from uniform periodicity. Because the uniformity of the periodicity is well established over the grating, then the present uncertainties in the ODs are all much reduced, as shown in the Figure 10.7.

The object of calibration for nanometrology measurement is not always directed to such a uniform artefact and comparison for SPM is becoming more and more important to nanoprobe users. One must also recognize that the scanning electron microscope is no longer used for calibration but exclusively for practical measurement and analysis. The measurement standards of each national measurement customer must be traceable to the NMIs and then recognized by global societies in a framework of agreement. NMIJ/AIST participated in this comparison by applying high-performance traceable AFM and has achieved excellent results. In this context, the 240 nm pitch standard microscale that is already supplied in Japan has proved

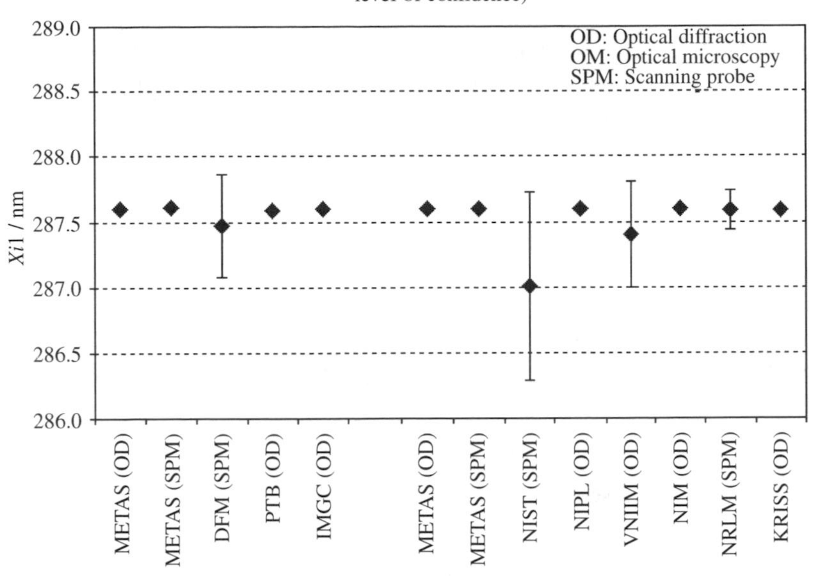

Figure 10.7 Results of international key comparison. From Website of Bureau International des Poids et Mesures (BIPM) key comparison database (KCDB), http://www.bipm.fr

acceptable internationally, as described earlier [10]. A similar comparison was also achieved in European countries [11].

With regard to depth direction scales, the Consulative Committee for Materials Quantification (CCQM) has measured the thickness of ultra-thin SiO_2 film on an Si substrate (measuring target thickness: 1.5 to 8 nm) in a pilot study during 2002–3 [12]. After this, a new working group was organized to deal with the field of surface and micro/nanoanalysis in fiscal year (FY) 2003. This strategy indicates that the project target is an internationally unexplored technical domain essential to the foundation of next-generation nanotechnology.

10.2 Nanomaterial Process Technology/Nanotechnology Material Metrology Project

This project is being conducted as a part of the nanotechnology programme in nanomaterial process technology, which aims to prepare a technical infrastructure for use in a wide range of nanotechnology industry by FY 2007. This will be achieved by developing process technologies suitable for fabricating nanostructures and their measurement technologies. In order to control nanostructures, it is very important to develop reliable measurement techniques for nanomaterials that run coherently from nanoscopic to macroscopic levels and are based on the common metrology standard. In addition, we require a universal standard for evaluating all aspects of nanomaterials, including their development, fabrication and application. To guarantee reliability and traceability of developed measurement methods, it is necessary to establish technical infrastructure for nanomaterials such as reference materials and measurement standards.

The research targets of the project are classified into the following four subthemes.

- measurement techniques for physical properties of fine particles and related standards;
- measurement methods and standard reference materials for nanopores;
- basic technology for measuring surface structures;
- measurement techniques for thermal properties of nanoscopic structures and related materials.

The needs, details, targets and results of the research into each subtheme are described in the next few sections.

10.2.1 Nanoparticle Mass/Diameter Measurement Technology

10.2.1.1 Particle Measurement Technology in Gas Phase

Nanoparticles are considered one of the key elements in nanotechnology. They can be building blocks of various nanoscopic structures; they are also important in the polymer, powder and biotechnology industries, as well as in environmental protection. Accurate measurement methods for physical properties of nanoparticles, such as size, mass and density, and standard materials related to these measurement methods are therefore important in these fields [13].

AIST has developed a method that enables highly accurate absolute measurements of mass for monodisperse particles suspended in the air. The principle of this method is similar to that of the Millikan method, in that both work by balancing the electrostatic and gravitational forces experienced by charged particles suspended between two plate electrodes. The unique feature of the AIST method is that the force balance is judged from the number of particles suspended after a certain holding time. In this way, it can be applied to particles as small as 100 nm, whereas the conventional Millikan method would be unusable due to Brownian motion of the particles. This new method is called the electrogravitational aerosol balance (EAB) method, and combined with an accurate particle density determination in which particles are immersed in density reference liquids, it gives a highly accurate particle diameter. The EAB is now used to develop particle size standards for the particle size traceability system in Japan.

AIST is trying to take the EAB method one step forward so that it can be applied to even smaller particles. The instrument in Figure 10.8 is currently under development and is called the aerosol particle mass analyser (APM). It uses centrifugal

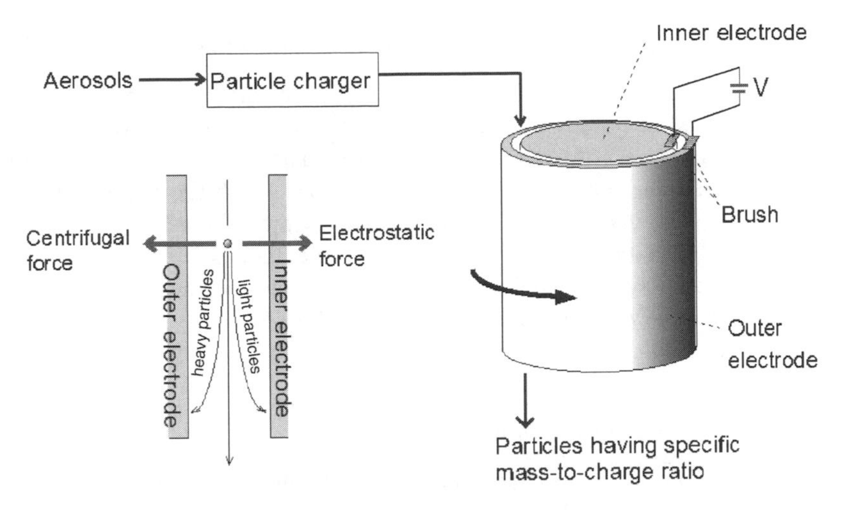

Figure 10.8 Principle of the aerosol particle mass analyser

force instead of the gravitational force used in the EAB. It works as a continuous classifier of particles according to their mass-to-charge ratio. Combined with a condensation particle counter used downstream of the APM, it can provide mass distribution of aerosol particles, as shown in Figure 10.9.

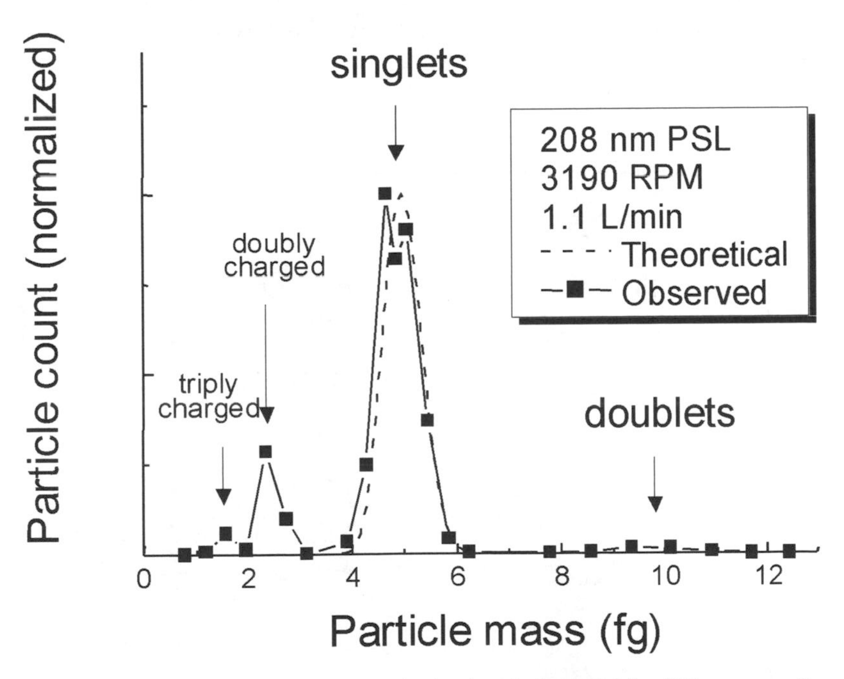

Figure 10.9 Mass distribution spectrum obtained with the APM for 280 nm monodisperse polystyrene latex (PSL) particles (mass about 5.0 fg) at 1.1 dm^3/min

10.2.1.2 Particle Measurement Technology in Liquid Phase

Currently under development is a technique for accurate diameter measurements of particles dispersed in liquids by using photon correlation spectroscopy [14]. The time correlation function of the light scattered from particles suspended in liquids is analysed to determine the diffusion coefficient, from which the particle diameter can be derived. The diameters are smaller than 100 nm. The adoption of a dual-correlator system, a high-power YAG laser as the light source, and a precise temperature control system has led to very accurate measurements. Also, nuclear magnetic resonance with pulsed field gradients (PEG-NMR) is being studied for particle size determination in the range 1–20 nm.

10.2.2 Nanopore Measurement Technology

Advanced nanoporosimetry is required for thin films such as low-*k* dielectrics used in next-generation semiconductors, high-sensitivity sensors, and nanocoatings for superior thermoresistance [15, 16]. AIST is developing a compact and easy-to-use positron lifetime spectrometer for use in small laboratories, both academic and industrial. This will take high-sensitivity nanoporosimetry based on positron annihilation and offer it to as many industrial users as possible.

The positron implanted into an insulator such as silica pairs with an electron to form positronium. Positronium annihilates after a short lifetime, the duration of which depends on the size of the nanopores (Figures 10.10 and 10.11). The nanopore size increases from 0.5 to 2.5 nm with additive concentration in the precursor solution, as shown in Figure 10.11. Thus far, AIST has assembled the

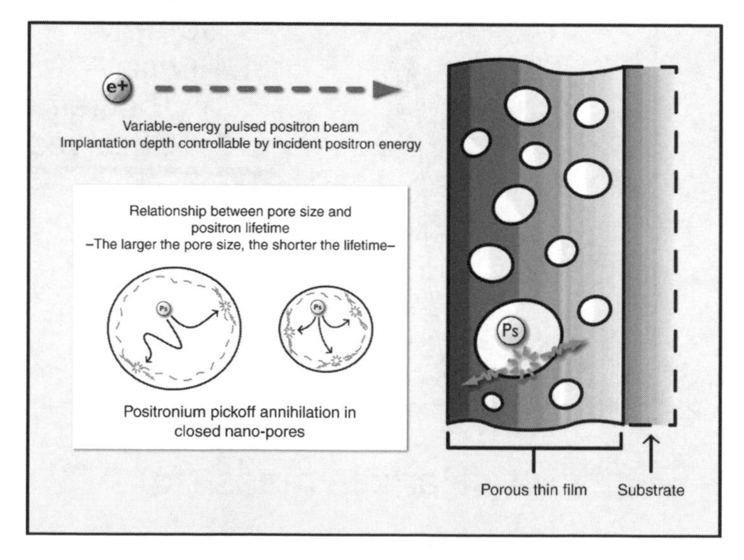

Figure 10.10 Principles of positron annihilation

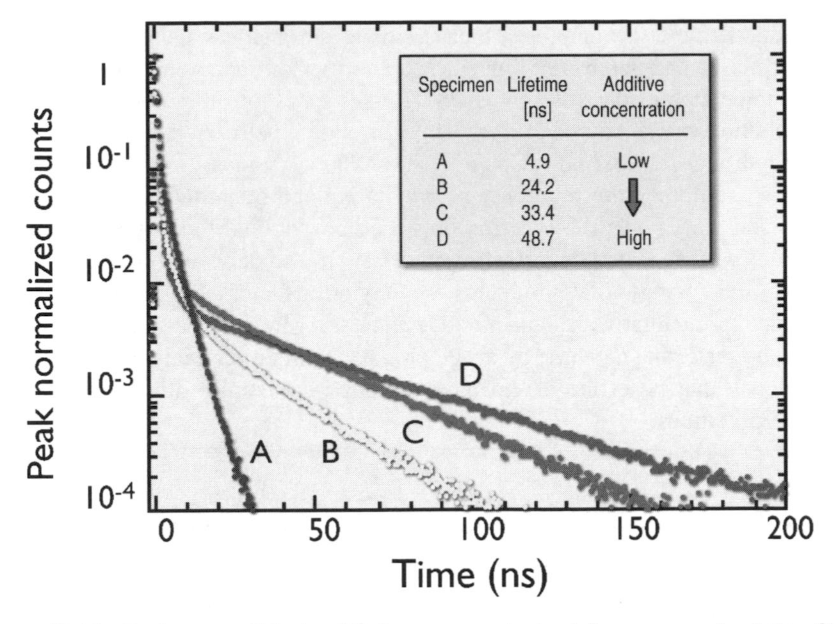

Figure 10.11 Positron annihilation lifetime curves obtained for porous sol-gel thin films with various porosities

spectrometer and confirmed the proper functioning of the beam transport and positron pulsing system composed of a chopper and buncher. The overall capability of the spectrometer is currently under examination with a radioactive positron source, sodium-22.

AIST is also engaged in developing standard samples for calibrating the spectrometer as well as ensuring confidence in the data obtained from it. Evaluation and optimization of nanopore structures of sol-gel porous silicate thin films is in progress so as to select materials for the standard samples. Positron nanoporosimetry is not only effective for developing porous films but also for evaluating gas barrier films that are used in flexible panel displays, reflection prevention films for optical components, and gas sensors.

10.2.3 Surface Structure Analysis Technology

Consider materials such as thin films, semiconductors and catalysts which have surface structures of nanometric order that are critical for their functions. Conventional photoelectron spectroscopy cannot perform accurate quantitative depth profiling surface analyses of these materials. However, it is expected that highly accurate non-destructive quantitative depth profiling surface analysis will be enabled by accurately defining the effective attenuation length and by including this data in the analysis [17, 18].

This subtheme is developing a measurement and analysis technique for use in non-destructive and depth profiling surface analyses of nanometric order based on X-ray photoelectron spectroscopy (XPS). It uses synchrotron radiation as the variable excitation energy source to clarify the correlation between the effective attenuation length and the energies of photoelectrons in substances.

It aims to improve the reliability of qualitative and quantitative analyses based on XPS and Auger electron spectroscopy, the conventional tools for effective elucidation of surface and boundary characteristics. It also deals with the construction of a spectrum database that eliminates as many error factors as possible and can be a reference for qualitative and quantitative analyses. Moreover, it will also construct an inelastic scattering database by developing a technique for analysing the inelastic backgrounds that pose one of the most serious problems in attempting accurate data interpretations.

10.2.4 Measuring Thermal Properties of Nanoscopic Structures

10.2.4.1 Technology for Thin Films and Boundaries

Highly reliable data on the thermal properties of thin films and the boundary thermal resistances are indispensable for thermal designs in the advanced technology fields [19, 20]. Typical applications are in the design of VLSI devices, phase-change optical and magneto-optical disks and phase-change non-volatile memory chips. AIST has developed a world-leading thin film thermal diffusivity measurement technology, picosecond thermoreflectance method. This detects surface temperature variations in the reflectance ratios that depend on temperature by heating the boundary between the thin film and the transparent substrate using a picosecond pulse laser (one picosecond is one-trillionth of a second). The technique is used to measure the thermal diffusivity of single-layer thin films with thicknesses below 1 μm and of the boundary thermal resistances of multilayer thin films. As the measurement targets of this technology have been limited to metallic films with thickness around 100 nm, this subtheme also aims at developing a technology that can measure the thermal diffusivity and thermal resistance of non-metallic thin film boundaries.

With the former picosecond thermoreflectance method, drift and output variation in the light intensity from the laser had been reflected directly in the amplitude signal, thereby leading to deterioration of the signal-to-noise ratio. However, we have already invented a detection method that is free from the effects of fluctuation in heat source light. We do this by observing the phase component instead of the conventionally observed amplitude component. This invention has drastically improved the signal quality. It is now possible to make highly reliable measurements of thermal diffusivity in metallic thin films having a thickness of 100 nm, allowing effective development of standard substances for thin film applications (Figure 10.12). In addition, this technology is now applicable to materials with low temperature coefficients of reflectivity that have previously proved difficult to measure, thereby expanding the usefulness of the technology.

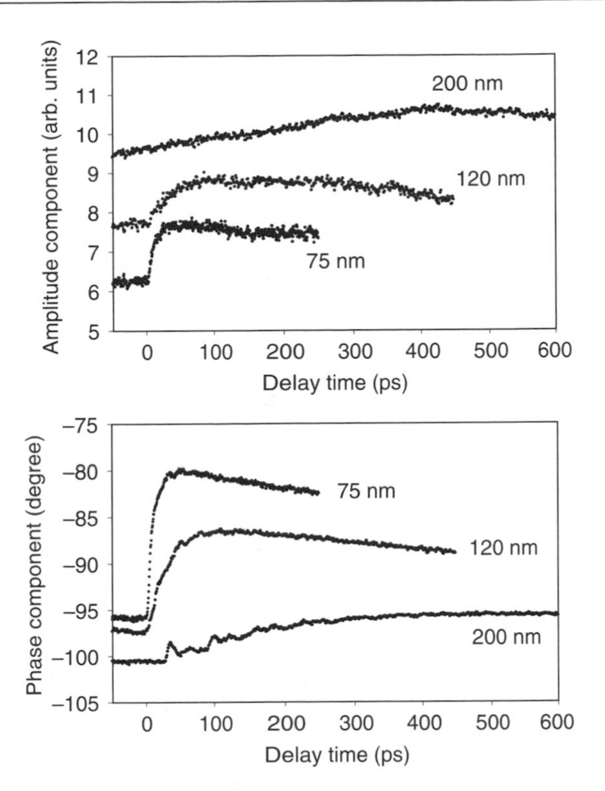

Figure 10.12 Thermoreflectance signals of molybdenum thin films synthesized on a glass substrate. The left-hand graph shows the signals obtained by the conventional method for detecting variations in the reflected light intensity. The right-hand graph shows the signals obtained by the phase detection method developed under this subtheme

Another problem with the previous picosecond thermoreflectance method is the restriction of the observable time span of measurements to about 1 ns after the picosecond pulse heating. This restriction was due to adjustment of the parallelism of the optical axis of the optical delay path. However, we have succeeded in extending the time span for observations of transient temperature changes to more than 10 ns by oscillating a pair of picosecond titanium-sapphire lasers synchronously and controlling the time interval of oscillation electrically (Figure 10.13). This method has allowed us to measure the thermal diffusivity and thin film boundary thermal resistances of non-metallic thin films, such as semiconductor thin films and oxide thin films as well as metallic thin films.

10.2.4.2 Thermal Expansion and Temperature-Dependent Changes in Optical Path Length

Low-expansion glasses and athermal glasses are important functional materials in the precision equipment, semiconductor and aerospace industries. Because these

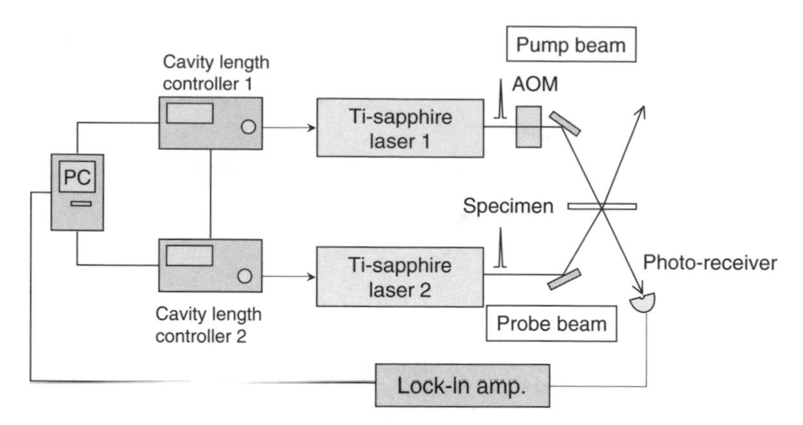

Figure 10.13 Block diagram of the newly developed thermophysical property measurement system (electrical delay system) based on the picosecond thermoreflectance method

glasses have to present very small thermal expansion and low temperature dependence in the *nL* product, we require technologies that offer highly accurate measurement and calibration functions to suit their specific needs. This subtheme is developing a technology for measuring the thermal expansion coefficients of solid materials and temperature changes in their optical path lengths (*nL* product) at room temperature at nanometric resolutions. This technique is based on a thermal expansion coefficient measurement technology in the possession of AIST that has the highest accuracy level in the world. With regard to the thermal expansion coefficient, it also aims to prepare a system for the supply of standard substances based on an absolute measurement and calibration technology.

10.3 Conclusion

Two national projects are currently trying to establish nanometrology standards at AIST. One is a project to produce nanoscales for lateral and depth directions, and to supply them as certified reference materials by fiscal year 2007. The other is a project to develop measurement methods for nanomaterials and carried out as part of the nanomaterial process technology. In addition to the certified reference materials dealt with in these projects, AIST is also planning to prepare others (a total of 250 substances by 2010) and to continue development by reviewing progress every year (www.nmij.jp/).

References

1. International Technology Roadmap for Semiconductors, 2001
2. S. Gonda, T. Doi, T. Kurosawa, Y. Tanimura, N. Hisata, T. Yamagishi, H. Fujimoto and H. Yukawa, *Review of Scientific Instruments*, **70**, 3362 (1999)

3. Y. Azuma, R. Tan, T. Fujimoto, I. Kojima, A. Shinozaki and M. Morita, in *Characterization and Metrology for ULSI Technology*, edited by D. G. Seiler *et al.*, AIP. Press, 2003, p. 337
4. A. Shinozaki, K. Arima, M. Morita, I. Kojima and Y. Azuma, *Analytical Sciences*, **19**, 1557 (2003)
5. T. Nishiguchi, Y. Morikawa, M. Kekura, M. Miyamoto, H. Nonaka and S. Ichimura, *Review of Scientific Instruments*, **73**(3), 1217 (2002)
6. T. Nishiguchi, H. Nonaka, S. Ichimura, Y. Morikawa, M. Kekura and M. Miyamoto, *Applied Physics Letters*, **81**, 2190 (2002)
7. K. Nakamura, S. Ichimura, A. Kurokawa, K. Koike, G. Inoue and T. Fukuda, *Journal of Vacuum Science and Technology*, **17**, 1275–1279 (1999): A. Kurokawa, K. Nakamura, S. Ichimura and D. W. Moon, *Applied Physics Letters*, **76**(4), 493 (2000)
8. BIPM homepage www.bipm.org
9. In Appendix B of the BIPM key comparison database, see the results of CCL-S1, supplementary comparison in length, dimensional meterology, pitch of gratings: 290 nm and 700 nm
10. I. Misumi, S. Gonda, T. Kurosawa, Y. Tanimura, N. Ochiai, J. Kitta, F. Kubota, M. Yamada, Y. Fujiwara, Y. Nakayama and K. Takamasu, *Measurement Science and Technology*, **14**, 2065 (2003)
11. R. Breil, T. Fries, J. Garnaes, J. Haycocks, D. Huser, J. Joergensen, W. Kautek, L. Koenders, N. Kofod, K. R. Koops, R. Korntner, B. Lindner, W. Mirande, A. Neubauer, J. Poltonen, G. B. Picotto, M. Pisani, H. Rothe, M. Sahre, M. Stedman and G. Wilkening, *Precision Engineering*, **26**, 296 (2002)
12. M. Seah *et al.*, *Surface Interface Analysis* (submitted)
13. K. Ehara, C. R. Hagwood and K. J. Coakley, *Journals of Aerosol Science*, **27**, 217–234 (1996)
14. K. Shimada, R. Nagahata, S. Kawabata, S. Matsuyama, T. Saito and S. Kinugasa, *Journals of Mass Spectrometry*, **38**, 948–954 (2003)
15. Y. Kobayashi, W. Zheng, T. B. Chang, K. Hirata, R. Suzuki, T. Ohdaira and K. Ito, *Journal of Applied Physics*, **91**(3), 1704–1706 (2002)
16. K. Ito, Y. Kobayashi, K. Hirata, H. Togashi, R. Suzuki and T. Ohdaira, *Radiation Physics and Chemistry*, **68**, 435–437 (2003)
17. N. Matsubayashi, T. Tanaka, M. Imamura, H. Shimada and T. Saito, *Analytical Sciences*, **17** (sup), 119–121 (2002)
18. M. Jo, *Surface Interface Analysis*, **35**, 729–737 (2003)
19. N. Taketoshi, T. Baba and A. Ono, *Measurement Science and Technology*, **12**, 2064–2073 (2001)
20. N. Taketoshi, T. Baba, E. Schaub and A. Ono, *Review Science Instruments*, **74**, 5226–5230 (2003)

Index

Nanotechnology: Global Strategies, Industry Trends and Applications Edited by J. Schulte
 ISBN: 0-470-85400-6 (HB)